Lecture Notes in Energy

Volume 46

Lecture Notes in Energy (LNE) is a series that reports on new developments in the study of energy: from science and engineering to the analysis of energy policy. The series' scope includes but is not limited to, renewable and green energy, nuclear, fossil fuels and carbon capture, energy systems, energy storage and harvesting, batteries and fuel cells, power systems, energy efficiency, energy in buildings, energy policy, as well as energy-related topics in economics, management and transportation. Books published in LNE are original and timely and bridge between advanced textbooks and the forefront of research. Readers of LNE include postgraduate students and non-specialist researchers wishing to gain an accessible introduction to a field of research as well as professionals and researchers with a need for an up-to-date reference book on a well-defined topic. The series publishes single- and multi-authored volumes as well as advanced textbooks.

Indexed in Scopus and EI Compendex The Springer Energy board welcomes your book proposal. Please get in touch with the series via Anthony Doyle, Executive Editor, Springer (anthony.doyle@springer.com)

Le Anh Nguyen Long · Dasom Lee

Energy Data

Imagining Future Energy Systems

Springer

Le Anh Nguyen Long
University of Twente
Enschede, The Netherlands

Dasom Lee
Korea Advanced Institute of Science and Technology
Daejeon, Korea (Republic of)

ISSN 2195-1284 ISSN 2195-1292 (electronic)
Lecture Notes in Energy
ISBN 978-3-032-19170-0 ISBN 978-3-032-19171-7 (eBook)
https://doi.org/10.1007/978-3-032-19171-7

This work was supported by Korea Advanced Institute of Science and Technology.

This Springer imprint is published by the registered company Springer Nature Switzerland AG
The registered company address is: Gewerbestrasse 11, 6330 Cham, Switzerland

To my parents—Elvie, Nguu, Rosa, and Salvatore—, to Francesco, Kai, Ale, and Eli, and of course to Dasom.

To my parents, Kabyoung and Bohyun. To my cats, Knightley and Pierre. And to my friend, Le Anh.

Preface

When we return home from work, we turn on the light in the living room, we check the fridge for food, we turn the television on, and we may even do some laundry. For many of us today, daily life and everyday practice is based on electricity, and we are certainly grateful for it. But we rarely think about who is checking how much electricity is being used by whom and when, for how long these data are being collected, how they are processed and analyzed, who benefits from these data and who is being marginalized, and what are the environmental consequences. Even as energy researchers, we (the authors of this book) do not often think about this carefully or critically enough because of the other distractions and obligations. Thinking hard on this issue is a luxury that working on this book has afforded us.

This book challenges this taken-for-granted system that we all live with. We identify the problems and opportunities with the existing energy system based on data that it uses. The current energy system collects energy consumption data, which is then analyzed and stored for an indefinite (and unknown to many) period of time. We question whether this practice is acceptable and just, what social and political problems it leads to, and who is liable for any mishaps. Then we invite you to join us in imagining a future where equity, sustainability, and privacy are respected and accounted for whilst taking inspiration from the work of colleagues who have developed frameworks for studying issues that are akin to the energy data paradox. These alternative futures are something we can only hope for in the current system.

The foundational yet may be not so obvious question that this book asks is: "Why do we often end up with systems that value efficiency and economic benefits above other social values such as sustainability, equity, and privacy?" This is connected to other important questions like, "Why are some values regarded higher than others? Who decides these values and why are we so forgiving to the technologies that neglect our social values that we claim we care about?" In attempting to address these questions, this book can also be construed as a message to the custodians of innovation, who claim that the core value lies in innovation and neglect social equity and environmental sustainability. Their tech-optimism fails to recognize how narrowly defined innovation often sidelines broader social commitments.

This book is an extension of our article "Data Donation: Using the gift relationship framework to address privacy and environmental issues of energy consumption data collection" published in *Energy Research & Social Science* in 2024. Initially, our aim in writing the book was to critically shine a light on problems associated with energy data. However, it became evident as writing progressed that future strategies regarding energy data must be included in the book if we are to seed ideas that can contribute to a present and a future that we all want to live in. This is partially because we are solution-oriented social scientists, but also because we believe that academic studies have the responsibility to draw hopeful and viable pictures that societies can aspire to and act towards. The U.S. Environmental Protection Agency once coined three Rs of climate change as "Reduce, Reuse, and Recycle" for people's daily climate action. This reflects an understanding of climate as limited to materiality and ignores the intangibles, like energy data. Therefore, as a rejoinder, our book is premised on a fourth R, *Reimagine,* which we believe can be the most powerful principle of the four.

We hope readers enjoy this book and take something from it: whether it be an awareness, a shift in how they perceive the systems around them, a conceptual tool, or a spark of insight. We hope this book kindles a conversation around data that we know is being collected but are so often, willfully and somewhat blissfully, ignored.

Enschede, The Netherlands — Le Anh Nguyen Long
Daejeon, Korea (Republic of) — Dasom Lee

About This Book

One of the overlooked aspects of the energy transition is data. Fine-grained and reliable energy consumption data are indispensable for smart grids, transactive energy systems, and renewable energy management. These technologies rely on high-frequency data to match energy demand and supply, enhance storage solutions, and promote sustainable consumption practices. However, the collection and use of massive datasets raise ethical and environmental concerns surrounding privacy, surveillance, and data justice. The increasing dependence on data in renewable energy systems thus embodies tensions between sustainability goals and data governance challenges. The prevailing discourses on energy data are thus highly polarized, between those who champion energy data and those who contest it. This book offers a middle ground: it invites a shift from data ownership to data stewardship. By drawing on frameworks from responsible innovation and problem shifting, this book interrogates not only what energy data are, but how they shape power relations, responsibilities, and notions of justice within the energy transition. Ultimately, the book aims to advance a vision for cultivating transparent, equitable, and ethically grounded practices that recognize energy data as both a technical resource and a socio-political construct.

Contents

About the Authors

Le Anh Nguyne Long is Assistant Professor of Public Administration at the University of Twente, Netherlands. Prior to this position, she was Researcher at the Center for Environmental Policy and Behavior (UC Davis) and at the Institute of Political Science (University of Münster). She examines energy and environmental policy innovation in the context of rapidly evolving socio-technical environments, which are prone to disruption. She has published in the *Journal of Cleaner Production, Big Data & Society, Energy Research & Social Science, Futures*, and *Global Environmental Politics* among others. She also has published two other books: *Co-creation and Smart Cities: Looking Beyond Technology*, and *Cities Against Democratic Backsliding: Evidence from Europe and Beyond*.

Dasom Lee is Associate Professor in the Graduate School of Science and Technology Policy at the Korea Advanced Institute of Science and Technology (KAIST). She worked as Assistant Professor at the University of Twente, the Netherlands, before joining KAIST. She specializes in renewable energy policy, sustainable transitions, and responsible innovation. She is an Associate Editor for *ACM Journal on Responsible Computing* and *Engaging Science, Technology, and Society*. She is also an Editorial Board Member for *Humanities and Social Science Communications*. Her interdisciplinary research has been published in journals such as *Environmental Innovation and Societal Transitions, Renewable and Sustainable Energy Reviews, Global Environmental Change*, and *Energy Research & Social Science* among others. For more information, visit www.dasomlee.com.

Abbreviations

AMI	Advanced Metering Infrastructure
BIPOC	Black, Indigenous, and People of Color
CCPA	California's Consumer Privacy Act
CEN	European Committee for Standardization
CENELEC	European Committee for Electrotechnical Standardization
DoS	Denial of Service
DSO	European Distribution System Operator
ESCO	Energy Service Company
ETSI	European Technical Standard Institute
EU	European Union
GDPR	General Data Protection Act
GPS	Global Positioning System
HSM	Hardware Security Modules
IT	Information Technology
LBNL	Lawrence Berkeley National Laboratory
MDMS	Meter Data Management Systems
NIS	EU's Network and Information Security
OT	Operational Technology
PII	Personally Identifiable Information
PIPA	South Korea Personal Information Protection Act
PKI	Public Key Infrastructure
RE	Renewable Energy
RI	Responsible Innovation
SEM	Structural Equation Modeling
SNOPUD	Snohomish County Public Utility District
UK	United Kingdom
US	United States of America
VEE	Validation, Editing, Estimation

Abbreviations

List of Figures

List of Tables

Part I
How Can Energy Data Be Understood?

Chapter 1
Introduction

1.1 Energy Data: Opportunities and Challenges

Renewable energy (RE) is argued to be one of the critical components of global efforts to mitigate climate change and presently failing attempts to keep the earth within its planetary boundaries. One of the overlooked aspects of the RE transition is *data*. In particular, nuanced and fine-grained data on the diverse sources of energy in the energy mix, how it is distributed, and patterns of use by individuals and households help smoothen the transition to RE. Accordingly, the focus of this book is on energy consumption data, *hereafter energy data,* that indicate how much electricity is used by households. Extensive household level energy consumption data is a lynchpin for promoting sustainable energy consumption and the use of RE (Lee and Hess 2021; Neumann and Mehlkop 2020; Pais-Magalhães et al. 2022).

This book focuses specifically on residential energy data rather than industrial or commercial energy consumption data. Residential energy data has been the most controversial in the literature (Lee and Hess 2021) as it can be used to predict individual behavior and lifestyle (Beckel et al. 2014; McDaniel and McLaughlin 2009). Additionally, commercial and industrial energy data tend to be considered a corporate secret, a possible indication of corporate production size, which can be useful information to its competitors. Therefore, the accessibility of commercial and industrial energy data is more of an issue compared to residential data. Finally, energy data policy implications for households and corporations are distinct and lead to distinct socio-technical dynamics. The decision to focus on households is rooted in the fact that individuals enjoy less protections. Because the general public has little knowledge about energy data, residential energy data is often used and shared without specific consent (Lee and Hess 2021). In this sense, residential energy data are more vulnerable to misuse or abuse. And yet they are critical for achieving an energy transition. Thus, energy data necessitates a critical interrogation and a re-imagination.

L. A. Nguyen Long and D. Lee, *Energy Data*, Lecture Notes in Energy 46,
https://doi.org/10.1007/978-3-032-19171-7_1

Box 1.1 *What is energy data? Energy data captures information about the generation, distribution, consumption, and environmental impact of energy in its various forms (e.g., electricity, gas, and photovoltaics). It encompasses a range of measurements and information from diverse sources like (smart) meters, building management systems, power plants, and industrial processes. Furthermore, energy data does not just refer to raw data that come from meters but data after they are cleaned, aggregated and processed*

Although the term *meter data* is commonly used in engineering and utility operations to describe the raw numerical values recorded by smart meters—typically time-stamped measurements of electricity consumption—*energy data* as a concept encompasses a much broader and politically layered terrain. Meter data is often treated as a technical artifact: a discrete, objective record used for billing, load forecasting, or operational efficiency. It is usually discussed in the context of system optimization, data transmission protocols, or IT architectures, with little consideration for its wider social or political implications

By contrast, *energy data* refers not only to what is collected by meters but also to the meanings, consequences, and governance structures that surround it. It includes who has access to the data, how it is used, who benefits from its insights, and who bears the risks. Importantly, the dominant "meaning" of energy data shifts from the point of production to distribution, storage, and use

Energy data is both an outcome of energy technology and social dynamics, thus they can bring about significant changes to existing technological infrastructures and processes. To elaborate, energy data is made accessible via an energy technology such as smart meters[1], at the same time the increasing of energy data also demands more efficient and more numerous data centers. As we will explore in this book, this is likely to change both the built and natural environment, in ways that may be consequential for our relationship with data, with our governing institutions, and with one another. Thus, engaging with energy data helps us better envision what kind of energy system we would like to have in the future.

This book puts the entire energy data system into view. In so doing, it invites the reader to examine energy data from diverse vantage points that can help reconcile the tensions between the different environmental and social benefits and challenges that energy consumption data constitutes. As the collection of data regarding the consumption of RE intensifies (Castro et al. 2024; Yang et al. 2023), tensions around data management and tradeoffs regarding data collection, sustainability, and privacy can be expected to become more apparent and to raise ethical and moral concerns around data justice and environmental responsibility. Balancing these concerns requires a mindset shift.

[1] Smart meters are interfaces that can frequently record energy consumption data remotely using wired or wireless communication technologies and reduce the need for manual monitoring or recording.

This book explores diverse pathways for sparking such a mindset shift. It is organized around three fundamental questions related to energy data:

1. How can energy data be understood?
2. Who gets what, when, and how in energy data?
3. How far can we take the insights developed here?

To offer new insights instead of regurgitations of existing knowledge, the starting point of this book is not the answers but rather the questions and the oftentimes contradictory responses to those questions from policy makers, industry players, non-governmental organizations, and the research community. To reconcile these diverse and fragmented responses, we begin by interrogating the different theoretical frameworks that can disrupt and enlarge current understandings and meanings of energy data. To that end we leverage insights and ideas from problem shifting and responsible innovation. Springing from this exploration, we outline the challenges related to the collection, use, and storage of energy consumption data. Each chapter also asks, "What can be done?" As a response for this, we present alternatives that illustrates how these concerns may be alleviated through data donation, the concept of consumption corridors to understand our consumption spectrum and to restrict endless collection of data, and imaginary future generations (IFGs) for future oriented visions.

1.2 The Importance of Energy Consumption Data and Challenges

Energy data is situated at the core of many energy system changes that scholars envision (see Fig. 1.1)—fairer and greater energy accessibility in the energy justice literature; more environmentally sustainable energy systems and greater RE adoption in the energy transition literature; better grid reliability in energy efficiency literature; and social challenges of new and emerging technologies in energy systems such as smart meters and smart grids[2]. These and other emerging technologies present a tremendous opportunity to meet current demand for high quality energy consumption data. However, collecting and storing these data is not without challenges. Challenges related to energy data are present at every phase of the energy data life-cycle. Individuals and environment are dually put at risk by the collection of massive sets of data, their ultimate fate remains unclear. Figure 1.1 further puts into view the diverse socio-technical infrastructures, which are built on and around energy data, and fuel current imaginaries around energy systems. All of these have important implications for our quality of life today and in the future. We consider four interlinked parameters of

[2] Smart grids are electricity network systems that use digital technologies to match supply and demand of energy consumption in real time (International Energy Agency, 2024).

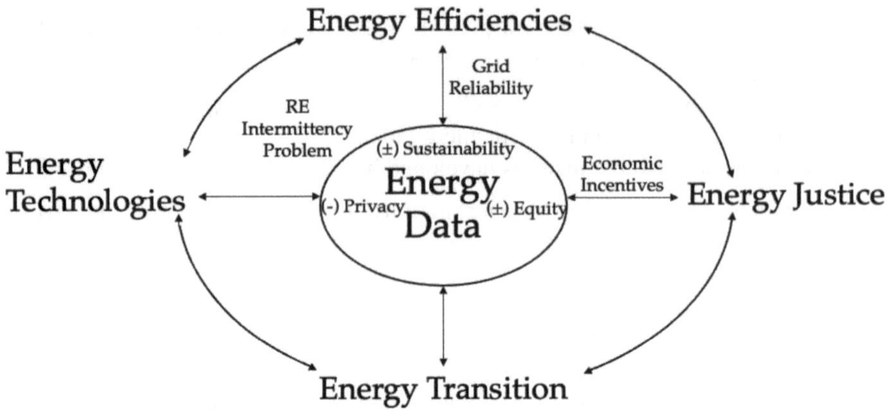

Fig. 1.1 Energy data interdependencies. *Caption: Energy data is coupled with other energy system elements, such as energy efficiencies, justice, technologies, and transitions. Through these interdependencies, they are also linked to other social goals like sustainability, privacy, and equity. The tendency to focus on the other energy system elements causes energy data to be overshadowed. This figure illustrates how energy data lies at the core of this assemblage*

energy systems. Namely, the efficiencies, transitions, technologies, and (in)justices that are informed and even perpetuated by energy data.

As is depicted in Fig. 1.1, *energy efficiency* is connected to energy data through grid technologies. More frequent sampling of energy data enables more accurate prediction of energy consumption. This can lead to greater energy efficiency, which is often what energy engineering researchers focus on (Avgerinou et al. 2017; Brown et al. 2007). This data allows two-way communication between producers and consumers, which enables consumers who are interested in sustainable energy practices to more easily convey their demands around renewable and sustainable energy (Lee et al. 2020). Efficiency suggests lower energy waste and increased alignment between energy supply and demand (Hopf et al. 2018; Nyborg and Røpke 2013). Given energy data's potential to unlock higher levels of energy efficiency, it can be seen as a pathway for ensuring energy security.

Energy data performs a lot of functions in the *energy transition*. The goal of energy transitions towards RE is undergirded by the coupling of energy policy and climate change policy. As such, energy and sustainability transitions become intertwinedcc 2024). To begin, energy consumption data is important in energy innovation for energy transitions. For example, it is needed to develop batteries that address RE intermittency (Ganesan et al. 2020; Yekini Suberu et al. 2014). Fluctuations in the supply of RE sources such as solar and wind render them unreliable. Several studies have tried to address RE intermittency through technological advances (Wang et al. 2013) and economic models (Gowrisankaran et al. 2016). To date, the most effective solution is energy storage systems that can be used during the periods of low sunlight or wind (Atawi et al. 2023; Yekini Suberu et al. 2014; Zsiborács et al. 2019). Energy consumption data is needed to measure the parameters of the required energy storage system. Additionally, without energy data it would be near impossible to deploy economic and control mechanisms needed to institute distributed energy systems through transactive energy[3]. Transactive energy requires granular energy consumption data to operate effectively (Ambrosio 2016; Lee et al. 2020). Consumption data makes transactive energy possible while giving consumers more control over their energy bills, opportunities to engage in more sustainable energy consumption practices, and participate in further RE developments (Chen and Liu 2017; Lee et al. 2020).

Reliability is another (invisible) feature of energy systems, on which our current social structures are built, as energy fuels many of those *technologies* (e.g., cars, phones, computers) that connect us to one another and make social interactions possible. Energy data can be used to ensure grid reliability, efficient energy use and management, cheaper energy pricing, and a higher likelihood of RE adoption (Chen and Liu 2017). Smart grids have facilitated demand-side management, RE management, and fault and infrastructure management. These practices ensure more sustainable and reliable energy systems, and require vast amount of energy consumption data (Lee and Hess 2021; Potdar et al. 2018).

Energy data has been associated with *energy justice*. With energy data, more efficient pricing of energy is possible. This can lead to more control given to customers in terms of their energy consumption (Chen and Liu 2017; Lee et al. 2020). Here the issue of energy poverty and energy affordability also come to fore. Some studies claim that this can lead to economic incentives such as lower energy bills because customers can run the dishwasher or washing machine during off-peak hours (Lee et al. 2020). However, there is a flip side. Others claim that lower income households are already saving as much money as possible; therefore, energy data collection only leads to exploiting lower income households' data rather than providing economic benefits. In other words, energy data comes with diverse energy justice implications, which puts the question of economic incentives into view.

Despite its importance, an in-depth discussion around energy data is missing. Many times, energy data is discussed as an afterthought, or tangential to what are considered to be more important topics such as energy policy analysis or RE transitions. In these discussions the equity, privacy, sustainability implications of current energy data regimes are minimally considered. And yet, as we hint at in the precluding discussion, these concerns are simultaneously addressed and stimulated during the collection, distribution, use and storage of energy data. Without careful reflection on these energy data externalities the likelihood of a "normal accident" or catastrophic failure by design related to energy data becomes an inevitability rather than a possibility (Perrow 1984), which jeopardizes the continued collection and use of

---

[3] Transactive energy systems respond to supply and demand of energy consumption in real time. It can include distributed energy resources and economic modeling of energy consumption/billing.

energy data. However, without good energy data, energy transitions will continue to be frustrated. Accordingly, this book seeks to enable a multi-faceted understanding of energy data, in its various forms with the hope of shining a light on a just and ethical energy data compromise and the good governance thereof.

1.3 Big Questions and Fragmented Answers About Energy Data

Energy data plays a critical role in shaping the direction and character of energy transitions: it informs decisions about infrastructure investment, enables or limits demand response programs, and influences the rollout of decentralized technologies like rooftop solar or battery storage (Avgerinou et al. 2017; Potdar et al. 2018; Yang et al. 2023). Moreover, the governance of energy data intersects with fundamental questions of energy justice—such as whether marginalized communities are subject to over-surveillance, whether these communities have equitable access to energy-saving programs, or whether data-driven transitions reproduce existing inequalities (Bietz et al. 2019; Lee et al. 2024). In this sense, energy data is not merely a technical input but a sociopolitical construct that requires critical examination beyond its engineering definitions. But energy data is so much more important than this.

Considered piecemeal, the social, technological, economic, political, and even cultural implications of energy data are hidden in plain sight. Putting together the pieces of the energy data puzzle allows us to consider our collective and personal relationships to energy data, and open the door to their transformation. Aligning with this, this book is an attempt to puzzle together the different understandings of this important yet often unseen player in energy systems as a way critically engage with the justice, responsibility, and democratic implications of energy justice. Accordingly, the book is organized around three questions concerning energy data.

In the first part of this book, we mine the social sciences literature on energy, environment, and society to tackle the question, *How can energy data be understood?* Part One thus includes this introduction, where we depart in Sect. 1.2 with the understanding of energy data as part of the complex solution to the trilemma represented by the energy transition, that of affordability, sustainability, and security. To deepen this discussion, the book follows with a (re)consideration of energy data that engages with the emerging critical discourse on digitalization and energy transitions. In that vein, Chap. 2 introduces and integrates responsible innovation (RI) and problem shifting. RI describes the processes, strategies, and norms of innovation, which should meet the public demand and social responsibility (Owen et al. 2012). However, responsibility is not the only social value that this book is organized around. Another core social value is justice (Bennear 2022; Droubi et al. 2022; Lee et al. 2024). Consequently, we introduce problem shifting as an important framework for considering the justice implications of energy data. Integrating problem shifting and RI, the book proposes in Chap. 2 (and later again in Chap. 4) that energy data is a technology that brings about social problems, which should be discussed and addressed.

As a rejoinder to the view on problems proposed under Chap. 2 and 3 shifts gears from energy data problems to solutions by examining energy governance. While RI situates energy data in the technology development phase, problem shifting invites questions of governance and policy. Accordingly, Chap. 3 points attention to the fragmented nature of energy data governance that is perpetuated by our fragmented approaches to energy data governance. To piece the energy data governance picture together, we leverage the literature on policy mixes and ecology of games. As a proof of concept, we apply these ideas when examining print media reporting on energy data and its governance. The chapter explores the different social goals which energy data is coupled with in the reporting and consider how the policy mix and its niches can enable problem shifting from one end of the policy landscape to another.

In Part Two energy data is viewed from a variety of vantage points to help you, the reader, see and understand how energy data can simultaneously contain multiple meanings for different stakeholders and how these understandings shape power relationships in energy systems. The coexistence of different energy data meanings and understandings reflect prevailing energy system power structures. Aligning with Luke's power theory, we understand power to operate on deep-seated preferences and beliefs that cause individuals to accept their situation as natural (Lukes 1974). As such, stakeholders may not fully grasp what is at stake for them or think that it is inappropriate to question decisions and actions on energy data production, sharing, and storage even when the decisions and actions run counter to their own interests (e.g., when energy producers sell consumption data to third parties).

Part Two engages with the question, *Who gets what, when, and how in energy data?* American political scientist called politics a process through which who gets what, when, and how is determined (Lasswell 1972). The result of these processes do more than distribute benefits and burdens (McDonnell and Elmore 1987), they also communicate to individuals where they lie within the socio-political hierarchy, they learn through politics and policy about their level of deservingness within these systems (A. Schneider and Ingram 1990; Schneider and Ingram 2005). This can also be seen in the case of energy data.

At any given moment, energy data is being produced, collected, stored, shared, and analyzed. These are the processes that lend meaning and generate different understandings of energy data, and they all happen contemporaneously. In this section of the book, we interrogate all of these processes through a careful application of the frameworks and ideas developed and imparted in Chaps. 2 and 3. For instance, we leverage RI to carefully consider energy data blind-spots. However, RI's broadness can complicate targeted analyses leading to watered down and abstract analyses of the innovation process (Blok and Lemmens 2015; Owen et al. 2012). Enter problem shifting, which we use to locate a technology or intervention at the sites from which a problem is shifted and the sites that it is shifted to, and in so-doing bring into the fold a justice and governance lens on energy data.

Each of the chapters in Part Two offer a different understanding of energy data. Chapter 4 defines it as a property, something that can be owned by energy producers, energy consumers, or both. Chapter 5 defines energy data as a commodity, which has a store of value which can be shared or traded. And finally, Chap. 6 sees it as capital, which puts the power of the means of production into the hands of those who claim to own it. Like capital, it can be invested to stimulate economic value creation, and it can also make other interests (e.g. data security, environmental sustainability) subservient to the goal of capital accumulation. Chapter 4 investigates energy data production, often via smart meters, and examines how it affects the problem owner (energy producers) and the problem receiver (electricity end-users). Situated in these microlevel dynamics, Chap. 4 considers how principles of energy data ownership, individual agency, and social justice can be upheld within energy data production process by accentuating the inclusive and responsive dimensions of RI. As such, Chap. 4 centers on a problem generated by energy data: on a loss of agency on the part of energy end users. Agency can be re-invigorated, we propose, by reimagining end-users relationship to energy data.

Chapter 4 explores problem shifting between actors, instead Chap. 5 explores problem shifting between domains via its institutions. Understanding institutions as prevailing rules and policies (Ostrom 1993), Chap. 5 investigates how energy data sharing solves energy institution problems at cost to privacy institutions. The meso-level analysis conducted in Chap. 5 leverages anticipatory and responsive principles of RI to examine the privacy problems related to energy data. The violation of privacy rules, it is argued, occurs primarily because companies can—with complete impunity in contemporary regulatory systems—treat data as property, something that can be owned and bartered. Moving the dial from owning to stewarding encourages ethical, responsible, and sustainable stewardship of energy data distribution systems, ensuring they are maintained, improved, and monitored in ways that prevent harm, promote societal well-being, and align technology with human values.

Finally, in Chap. 6 problem shifting sheds light on dynamics between communities triggered by energy data storage. The energy data collected using smart meters also demand larger and more data centers for data storage, which links to studies on energy decentralization (use of small modular reactors or local energy to promote co-location and to reduce energy loss), energy efficiency (more efficient use of data center energy), and energy justice (where to place data centers). With the increasing amount of energy data being stored every day in data centers sustainability challenges arise, shifting problems from one community to a different geography or time.

Harvesting insights from Chaps. 4–6, the chapters in the third part of the book address the question. *How far can we take the insights developed here?* In addressing this question, we consider if different imaginaries about energy data may transform how we regulate and govern data. Chapter 7 tackles this question from the vantage point of policy and politics. It reveals how equity, privacy, and sustainability interdependencies are baked into energy data systems through the energy policy mix. The concept of policy mix assumes that policy domains overlap and have externalities driven by problem shifting and other dynamics. Chapter 7 unpacks the sometimes complementary yet other times contradictory relationships among energy data equity, privacy, and sustainability considerations. This exercise is needed to imagine the future energy systems where many of these problems are addressed—if not solved—through regulations, new system designs, and new and emerging technologies.

The book concludes with Chap. 8, which returns to the question why energy data is important and why it should be placed at the core of many energy literatures. In so doing, this chapter sets the research agenda for future studies. This concluding chapter discusses the multifaceted dimensions of energy data, considers how insights on energy data can be applied to other data systems and ends with a note of encouragement of transdisciplinary research around energy data.

1.4 Navigating This Book

This book is both an appeal and a challenge. The appeal is for the reader to slow down and (re)consider the energy system and its evolution through the lens of energy justice. The challenge is to be willing to consider energy data as more than an object or a means to an end but also as a *technology* that has its own power and imaginaries. To meet the appeal and diminish the challenge the chapters of this book are written to stand alone. Readers who are interested in understanding the specific facets or processes related to energy data can read the chapters that are relevant to their interest. The abstracts and keywords have been included to help readers determine the alignment between their interests in energy data and the chapters that we have prepared.

Of course, readers can expect to benefit most by reading the entire book. The eight chapters of this book are divided into three parts, which as we explain above is organized around a different energy data question. Essentially, Part One challenges readers to reconsider how they understand energy data by seeing it as a vehicle for problem shifting, which therefore must be treated with responsibility (Chap. 2) and also by seeing it current governance through a misunderstood policy mix as a structure that reinforces problem shifting (Chap. 3). In Part Two, energy data is interrogated in terms of responsibility and problems prevalent to energy data processes like production (Chap. 4), storage (Chap. 5), and sharing (Chap. 6). In so doing, the book takes a deep dive into the three social issues at its core: privacy, social marginalization (justice), and sustainability (See Table 1.1).

Table 1.1 Understanding energy data through its production, distribution, and storage

Chapter	Problem shifting between	Responsible Innovation considerations:
Chapter 4	Energy producers and consumers (micro—level dynamics)	Reveals inequity in Energy Data Production by: • Considering the marginalization of end-users whose data rights are not recognized or respected (Inclusion) • Considering the choice to respond to the needs of wealthy end-users and energy providers rather than poor energy end-users (Responsiveness)
Chapter 5	Energy and privacy domains (Meso—level dynamics)	Reveals privacy concerns in Energy Data Sharing by: • Considering the possible future security risks of smart grids (Anticipation) • Considering the failure to respond to the very real privacy needs of energy end users (Responsiveness)
Chapter 6	Energy data users and communities near data centers and future communities (macro—level dynamics)	Reveals sustainability concerns in Energy Data Storage by: • Considering the consequences for future generations of current consumption (Anticipation) • Considering the needs of communities living near data centers or whose drinking water is put at risk by data centers (Inclusion)

Finally, Part Three presents alternative imaginaries of energy data (Chap. 7) to set the stage for a reimagining of it, its governance, and research around it (Chap. 8). We summarize these ideas in Table 1.2 below.

The lens in these chapters moves from problems to power, and the reader is presented with a consideration of how the disruptive potential of socio-technical transitions like those that the energy system is experiencing go beyond technology and touch upon culture. The policy and research agenda put forward in Chap. 8 serves both as a conclusion to this book and an invitation to continue engaging with these ideas in order to develop new knowledge and understandings of energy data.

Table 1.2 How do you solve a problem like energy data?

	Challenge	Alternative pathway	Ethos	Governance Strategies
Micro	Inequality	Data donation by end-user	Gift relationship	Inclusive and participatory governance
Meso	Privacy	Consumption Corridors	A good life	Reimagining limits on energy data use as a means to pursue a good life rather than curtail independence
Macro	Sustainability challenges	Imaginary Future Generations	Seventh generation principle	Anticipatory Governance processes

References

Ambrosio R (2016) Transactive energy systems. IEEE Electrification Mag 4(4):4–7

Atawi IE, Al-Shetwi AQ, Magableh AM, Albalawi OH (2023) Recent advances in hybrid energy storage system integrated renewable power generation: configuration, control, applications, and future directions. Batteries 9(1), Article 1. https://doi.org/10.3390/batteries9010029

Avgerinou M, Bertoldi P, Castellazzi L (2017) Trends in data centre energy consumption under the European code of conduct for data Centre energy efficiency. Energies 10(10), Article 10. https://doi.org/10.3390/en10101470

Beckel C, Sadamori L, Staake T, Santini S (2014) Revealing household characteristics from smart meter data. Energy 78:397–410. https://doi.org/10.1016/j.energy.2014.10.025

Bennear LS (2022) Energy justice, decarbonization, and the clean energy transformation. Annu Rev Resour Econ 14(1):647–668. https://doi.org/10.1146/annurev-resource-111920-022328

Bietz M, Patrick K, Bloss C (2019) Data donation as a model for citizen science health research. Citiz Sci: Theory Pract 4(1):6. https://doi.org/10.5334/cstp.178

Blok V, Lemmens P (2015) The emerging concept of responsible innovation. Three reasons why it is questionable and calls for a radical transformation of the concept of innovation. Springer International Publishing. https://doi.org/10.1007/978-3-319-17308-5_2

Brown R, Masanet E, Nordman B, Tschudi B, Shehabi A, Stanley J, Koomey J, Sartor D, Chan P, Loper J, Capana S, Hedman B, Duff R, Haines E, Sass D, Fanara A (2007) Report to congress on server and data center energy efficiency: public law 109-431 (No. LBNL-363E, 929723; Issue LBNL-363E, 929723, p LBNL-363E, 929723). https://doi.org/10.2172/929723

Castro V, Georgiou M, Jackson T, Hodgkinson IR, Jackson L, Lockwood S (2024) Digital data demand and renewable energy limits: forecasting the impacts on global electricity supply and sustainability. Energy Policy 195:114404. https://doi.org/10.1016/j.enpol.2024.114404

Chen S, Liu CC (2017) From demand response to transactive energy: state of the art. J Mod Power Syst Clean Energy 5(1):10–19. https://doi.org/10.1007/s40565-016-0256-x

Droubi S, Heffron RJ, McCauley D (2022) A critical review of energy democracy: a failure to deliver justice? Energy Res Soc Sci 86:102444. https://doi.org/10.1016/j.erss.2021.102444

Ganesan S, Subramaniam U, Ghodke AA, Elavarasan RM, Raju K, Bhaskar MS (2020) Investigation on sizing of voltage source for a battery energy storage system in microgrid with renewable energy sources. IEEE Access 8:188861–188874. https://doi.org/10.1109/ACCESS.2020.3030729

Geels FW (2019) Socio-technical transitions to sustainability: a review of criticisms and elaborations of the multi-level perspective. Curr Opin Environ Sustain 39:187–201. https://doi.org/10.1016/j.cosust.2019.06.009

Gowrisankaran G, Reynolds SS, Samano M (2016) Intermittency and the value of renewable energy. J Polit Econ 124(4):1187–1234. https://doi.org/10.1086/686733

Hopf K, Sodenkamp M, Staake T (2018) Enhancing energy efficiency in the residential sector with smart meter data analytics. Electron Mark 28(4):453–473. https://doi.org/10.1007/s12525-018-0290-9

International Energy Agency (2024) Smart grids—IEA. https://www.iea.org/energy-system/electricity/smart-grids

Lasswell HD (1972) Politics: who gets what, when, how (11. print). Meridian Books

Lee D, Hess DJ (2021) Data privacy and residential smart meters: comparative analysis and harmonization potential. Util Policy 70:101188

Lee D, Hess DJ, Neema H (2020) The challenges of implementing transactive energy: a comparative analysis of experimental projects. Electr J 33(10):106865

Lee D, Long LAN, Jansma SR (2024) Data donation: using the gift relationship framework to address privacy and environmental issues of energy consumption data collection. Energy Res Soc Sci 114:103596. https://doi.org/10.1016/j.erss.2024.103596

Lee J-H, Woo J (2020) Green new deal policy of South Korea: policy innovation for a sustainability transition. Sustainability 12(23), Article 23. https://doi.org/10.3390/su122310191

Lukes S (1974) Power: a radical view. Macmillan. https://books.google.nl/books?id=UB2qzw EACAAJ
McDaniel P, McLaughlin S (2009) Security and privacy challenges in the smart grid. IEEE Secur Priv 7(3):75–77. https://doi.org/10.1109/MSP.2009.76
McDonnell LM, Elmore RF (1987) Getting the job done: alternative policy instruments. Educ Eval Policy Anal 9(2):133–152. https://doi.org/10.3102/01623737009002133
Neumann R, Mehlkop G (2020) Framing electricity plan choices to enhance green energy usage: a choice experiment with panel data from Germany. Energy Res Soc Sci 70:101741. https://doi.org/10.1016/j.erss.2020.101741
Nyborg S, Røpke I (2013) Constructing users in the smart grid—insights from the Danish eFlex project. Energ Effi 6(4):655–670. https://doi.org/10.1007/s12053-013-9210-1
Ostrom E (1993) Design principles in long-enduring irrigation institutions. Water Resour Res 29(7):1907–1912. https://doi.org/10.1029/92WR02991
Owen R, Macnaghten P, Stilgoe J (2012) Responsible research and innovation: from science in society to science for society, with society. Sci Public Policy 39(6):751–760. https://doi.org/10.1093/scipol/scs093
Pais-Magalhães V, Moutinho V, Robaina M (2022) Is an ageing population impacting energy use in the European Union? Drivers, lifestyles, and consumption patterns of elderly households. Energy Res Soc Sci 85:102443. https://doi.org/10.1016/j.erss.2021.102443
Perrow C (1984) Normal accidents: living with high-risk technologies. Basic Books
Potdar V, Chandan A, Batool S, Patel N (2018) Big energy data management for smart grids—issues, challenges and recent developments. In Mahmood Z (ed) Smart cities: development and governance frameworks. Springer International Publishing, pp. 177–205. https://doi.org/10.1007/978-3-319-76669-0_8
Restructuring energy systems for sustainability? Energy transition policy in the Netherlands—ScienceDirect (2024). https://www.sciencedirect.com/science/article/pii/S030142150800308X?casa_token=YbqKZ7KikbkAAAAA:gwdHjvYcnpJg5UeMHogQkOgoQZpLFjfhItPt8yRuLgadnEObNnHaSSr3Z-UbJWsb66Ir8zF3jMI
Schneider A, Ingram H (1990) Behavioral assumptions of policy tools. J Polit 52(2):510–529. https://doi.org/10.2307/2131904
Schneider AL, Ingram HM (eds) (2005) Deserving and entitled: social constructions and public policy. State University of New York
Wang L, Zhang H, Chen D (2013) Intermittency indexes for renewable energy resources. IEEE Power Energy Soc Gen Meet 2013:1–5. https://doi.org/10.1109/PESMG.2013.6672651
Yang C, Bu S, Fan Y, Wan WX, Wang R, Foley A (2023) Data-driven prediction and evaluation on future impact of energy transition policies in smart regions. Appl Energy 332:120523. https://doi.org/10.1016/j.apenergy.2022.120523
Yekini Suberu M, Wazir Mustafa M, Bashir N (2014) Energy storage systems for renewable energy power sector integration and mitigation of intermittency. Renew Sustain Energy Rev 35:499–514. https://doi.org/10.1016/j.rser.2014.04.009
Zsiborács H, Baranyai NH, Vincze A, Zentkó L, Birkner Z, Máté K, Pintér G (2019) Intermittent renewable energy sources: the role of energy storage in the European power system of 2040. Electronics 8(7), Article 7. https://doi.org/10.3390/electronics8070729

Lukes S (1974) Power: a radical view. Macmillan. https://books.google.com/books?id=B2qw [illegible] EACAAJ

McDaniel P, McLaughlin S (2009) Security and privacy challenges in the smart grid. IEEE Secur Priv 7(3):75–77. https://doi.org/10.1109/MSP.2009.76

McDonnell LM, Elmore RF (1987) Getting the job done: alternative policy instruments. Educ Eval Policy Anal 9(2):133–152. https://doi.org/10.3102/01623737009002133

[illegible]

Chapter 2
Responsibility, Justice, and Energy Data

2.1 Theoretically Framing Energy Data

Research in innovation studies and management has provided a key starting point for scholarly discussions of meter data and the wider energy transition. Much of this work has drawn on sociotechnical transitions frameworks, which examine how large-scale technological changes—such as the shift from fossil fuels to low-carbon energy systems—unfold through the interplay of technological innovation, market structures, policy regimes, and societal practices (Geels 2014, 2019; Geels et al. 2017; Hess 2024; Martinez 2022; Raven 2024). Within this perspective, smart meters and the data they generate are often analyzed in terms of their capacity to support system optimization, integrate renewable energy, and enable new market mechanisms. Meter data, valued primarily for its precision in billing and demand forecasting, has also figured prominently in research on energy poverty, by revealing patterns of household consumption, and in energy economics, where it informs price modeling and the design of dynamic tariffs (Doostizadeh and Ghasemi 2012; Hatch and Graff 2024; Mahler et al. 2022; Ye et al. 2023).

This book approaches energy data from a different vantage point. Rather than treating energy data simply as an informational resource for economic analysis, technology (re)development, or policy design, it conceptualizes energy data as a technological object embedded in social relations and power structures. Adopting a science and technology studies (STS) lens foregrounds how data practices are shaped by human values, institutional arrangements, and contested notions of privacy and justice. From this perspective, energy data is not merely a neutral input to markets or infrastructure but part of a socio-technical assemblage that influences how energy is produced, distributed, and governed. Such an approach allows for a more critical interrogation of how data infrastructures (re)distribute and maintain authority, how they create new forms of dependency or exclusion, and how they might be reconfigured to support more equitable and democratic energy futures.

L. A. Nguyen Long and D. Lee, *Energy Data*, Lecture Notes in Energy 46,
https://doi.org/10.1007/978-3-032-19171-7_2

Among the various theoretical perspectives developed within STS Responsible Innovation (RI) stands out as particularly relevant for examining energy data. RI is explicitly concerned with anticipating and governing the societal consequences of technological change, seeking to align innovation processes with broadly shared values such as sustainability, inclusiveness, and fairness. Although other STS approaches, such as actor–network theory or social construction of technology, provide valuable descriptive and analytical insights, they do not always offer normative guidance for how innovation should proceed. RI, in contrast, combines critical analysis with practical orientation; it can be used to examine not only how energy data systems emerge and gain legitimacy but also can specify mechanisms—like stakeholder deliberation, anticipatory scenario building, and iterative evaluation—that can steer their development toward more socially desirable outcomes. This normative and procedural focus makes RI a particularly fitting framework for interrogating and shaping the evolving landscape of energy data. In other words, RI is a theoretical and normative framework that critically interrogates and challenges the often-unquestioned assumption that all innovation is inherently beneficial or necessary. Rather than accepting technological progress as an unqualified good, RI invites a systematic examination of the motivations, purposes, and potential consequences of innovation processes.

Despite the significant scholarly impact that RI theory has had on innovation, it has also been subject to criticism for being overly abstract and difficult to operationalize in practice (Burget et al. 2017; Genus and Iskandarova 2018; Genus and Stirling 2018). Because energy data is a severely under researched topic, we need a theoretical framework that is holistic but also has policy implications that lead to outcomes. To advance a more policy-relevant and action-oriented understanding of RI in the context of energy data, this chapter argues that theorizing around problem shifting can be a valuable complement to RI that offers concrete analytical tools and regulatory pathways to identify, anticipate, and address the displacement of environmental and social harms across domains.

2.2 Responsible Innovation

A central aim of RI is to challenge the notion that technological advancement should be pursued solely for its own sake—a perspective that is often, though not exclusively, found within engineering and related disciplines. Rather than treating innovation as inherently beneficial or neutral, RI calls for a more critical and deliberative approach that considers the broader societal context in which innovation occurs. This includes careful reflection on potential social, ethical, and environmental consequences, as well as ongoing engagement with diverse stakeholders to ensure that technological developments are aligned with collective values and public interest (Macnaghten 2020). Therefore, RI seeks to interrupt the momentum of unreflective or path-dependent technological development by posing fundamental questions about why particular innovations are pursued, who benefits from them, and what social, ethical, or environmental implications they may entail (Stilgoe et al. 2013). In doing so, RI promotes a more deliberate, inclusive, and value-sensitive approach to innovation that foregrounds responsibility, reflexivity, and societal alignment (Owen and Pansera 2019).

RI is interested in not only the deployment and use of technologies (i.e., the responsibility that is shaped by consumers and other end users) but also production of technologies (i.e., conceptualization, production, and institutionalization of technologies), which includes (and is certainly not limited to) questions such as how to scientists and engineers perceive their responsibility and what criteria constitutes scientific excellence (Pandza and Ellwood 2013; Pavie et al. 2014). Furthermore, RI also introduces the future strategies for technology and innovation into the picture, what kind of futures we imagine, what windows of opportunities do we need to create, and what kind of spatial and temporal differences and distinctions we must consider (Owen et al. 2013).

RI derives power from the systemic analysis that it proposes. It does not leave the responsibility—or the irresponsibility – of science and technology to individuals' morality, but rather it highlights how social and political systems of power, capital, and institutions structure the conditions under which innovation occurs and expands (Stilgoe 2013). RI shifts the focus from isolated ethical and political—power related—decision making to the broader sociotechnical arrangements that shape research agendas, funding priorities, and both public and private governance mechanisms. By calling for responsibility to be embedded into these structural dimensions, RI enables a more multidimensional and systemic approach to innovation rather than simply blaming or praising a single or a group of individuals for their success or failures. In essence, RI recognizes the entanglement of technology with public values, democratic legitimacy, and long-term societal consequences.

It is helpful to compare Ulrich Beck's concept of organized irresponsibility to RI (Beck 1992). Organized irresponsibility discusses how the responsibility of innovation and environmental hazards are diffused across institutions, and how this consequentially leads to fragmented and distributed responsibility (Beck 1997; Gumbert 2022a). He attributes this to bureaucratic structures, legal frameworks, and technical rationality (Galantino 2022; Gumbert 2022b). RI is distinguished from organized irresponsibility in that it calls for systemic accountability instead of systemic avoidance, proactive regulations and prevention rather than reactive, and transparency (highlighted by reflexivity and responsiveness) instead of opacity. In other words, if organized irresponsibility is an account of bureaucratic, institutionalized, and systemic failure, RI can be seen as a corrective framework that leads to greater moral engagement.

Many studies used RI framework to discuss the responsible innovation of new and emerging technologies, which are particularly vulnerable to falling in the trap of innovation for the sake of innovation for two reasons. First, there are few regulations in place to manage the technologies, which leaves technology developers to their own devices during the innovation process. Small unmanned aerial vehicles (sUAVs) is a great example of this—initially introduced as toys, regulators saw no need to regulate small sized drones. Now with many of them increasing in size and with camera and voice recorders attached, regulators are finally introducing relevant safety regulations. However, because the owners of sUAVs were not managed or controlled from the very beginning, the entire process of regulating sUAVs is left to the goodwill of citizens, who are expected to voluntarily take courses and register their sUAVs (Lee et al. 2022). Second, companies race with each other to be the first to "successfully" deploy the technology to gain an upper hand in the market and to attach their name to the technology in question. Therefore, processes that may cause latency – anticipation, reflexivity—are often neglected in the initial stages of innovation.

Jack Stilgoe, one of the leading figures of RI, focuses on automated vehicles and how RI's multi-stakeholder, multi-dimensional accountability and accessibility is ignored in this new transportation technology (Cohen et al. 2020; Stilgoe 2021; Stilgoe and Mladenović 2022), Other scholars focus on nanotechnology (Amutha et al. 2024; Cummings et al. 2021; Kokotovich et al. 2021; Wiek et al. 2016), digital health (Alami et al. 2024; Eyre et al. 2020; Iakovleva et al. 2019; Naughton et al. 2023), and artificial intelligence (Buhmann and Fieseler 2021; Herrmann 2023; Stahl 2022). The influence that RI carries is evidenced in the European Union's mainstreaming of responsible research and innovation principles across its various innovation research funding mechanisms like the Horizon Europe framework (van den Hoven 2022).

Despite the success and influence of RI and its importance in shaping a space where actors and institutions are held responsible, which we strongly support, the theory also has come under several criticisms. The most common and agreed upon critique is that RI is too abstract and lacks concrete mechanisms for operationalization, particularly in terms of translation into policy instruments (Owen and Pansera 2019). It is considered as an *ideal* that does not have clear pathways for implementation (Blok and Lemmens 2015; van de Poel and Sand 2021). This gap between principle and practice undermines its transformative potential especially within sociotechnical systems, where institutional inertia, power asymmetries, and conflicting interests are the norm.

Furthermore, de Hoop et al. (2016) argued that RI underestimates material constraints such as infrastructure, water, and competing land uses, fails to address the social and economic costs of innovation, lacks clear demarcation of responsibilities (which can fall into the trap of organized irresponsibility), and that it does not sufficiently consider "non-innovation" as a valid and responsible outcome. Similarly, Blok and Lemmens (2015) claim that RI places too much emphasis on process rather than outcomes, does not account for tension between innovation, which inherently carries risk, uncertainty, disruption and responsibility and creates a moral overload, where unrealistic demands are expected from innovators leading to superficial compliance or strategic appropriation.

Because of these limitations, RI on its own does not over the level of practical specificity or operational clarity needed for the kind of analytical and normative work pursued in this book. Although it provides an essential ethical compass and a rich language for examining social dimensions of technological change, RI often lacks concrete mechanisms for identifying how harms are distributed or for determining who should bear responsibility for preventing those harms and when innovation generates unintended consequences.

To address these theoretical weaknesses, we place RI in dialogue with the framework of problem shifting, which offers fine-grained account of how environmental, social, and institutional burdens migrate across actors, sectors, and temporal scales. By seeing a problem not just in terms of where it lands but also in terms of where it came from, the attribution of responsibility can be more accurately performed and governed. By merging these two perspectives, we seek to construct a more comprehensive and actionable foundation for understanding and governing energy data.

2.3 Problem Shifting

Problem shifting, which also was a rather simple and may be even an abstract idea that described the movements—or shifts—of environmental damage and hazards to other locations, temporalities, and dimensions (Wood Hansen and Van Den Bergh 2023), became much more policy oriented when it was applied to international environmental law. Kim and van Asselt (2016) argued that through international organizations and international conventions, provisions that prevent or at least take account of potential problem shifting can be devised. Although they bring attention to the current lack of discussion of problem shifting in public international law and governance, they also create sufficient space for international organizations and national governments to engage with such discussions by suggesting the following: "establishing an integrated management plan for the whole Earth System," (Kim and van Asselt 2016, p 493), assessment of problem shifting led by the Intergovernmental Panel on Climate Change, impact assessments completed by treaty bodies, inter-treaty cooperation, and a clearly defined and unified goal for all international regulatory regimes.

Problem shifting as a concept is not new. Several studies have attempted to map out environmental problem shifting of energy such as solar (van den Bergh et al. 2015) and gasoline (Yang et al. 2012). The concept considers how solving or attempting to solve an environmental problem can shift the problem to other geographies or time spans in the form of other environmental or social consequences. A good and commonly known example of problem shifting include electric vehicles as a solution to emissions while requiring inputs like lithium batteries that lead to intense metal mining and related labor issues (Boateng and Klopp 2024; Egbue 2012)—an attempt to build a sustainable mode of transport led to other social (labor) and environmental (rare metal extraction) challenges. A historical example of problem shifting can be seen in the invention of plastic, which was initially developed in the mid-19th century as a substitute for ivory in billiard balls. Although this innovation aimed to provide a more accessible material and potentially save elephants, it inadvertently set the stage for a global environmental crisis in the future. Over time, the widespread production and use of plastic led to persistent pollution, culminating in today's challenges of micro- and nano plastics infiltrating ecosystems, food chains, and even human bodies, which illustrates how solving one environmental problem can give rise to new, complex harms.

Following from these examples, it is evident that problem shifting is spatially and temporally dynamic. In this regard, the trajectory envisioned by problem shifting analysis aligns closely with the objectives of RI, touching on dimensions anticipation and reflexivity. Furthermore, problem shifting travels across different domains – an environmental challenge addressed can lead to a social problem or one area of environmental challenge solved can lead to another environmental challenge. For example, biofuels were initially promoted as a climate friendly alternative to fossil fuels (i.e., climate domain) but it led to land use conflicts (i.e., social domain), food and fuel debate (i.e., economic and social domains), and deforestation and biodiversity loss (i.e., ecological domain) (Gao et al. 2011; German et al. 2011). In this sense, problem shifting has three dimensions: space, temporality, and domains.

Problem shifting as a concept became much more empirical since Kim began systematically collecting global cases and publicly sharing the data through an open-access platform (https://www.problemshifting.directory/case/). This empirical turn not only enhanced the analytical utility of the concept but also enabled comparative analysis across sectors and geographies, grounding theoretical insights in real-world observations. Despite this progress, problem shifting remains difficult to identify. Its complexity and the imperceptibility of its harmful impacts – especially when these are displaced far into the future or onto geographies that receive little attention from the public, media, or government—make it hard to recognize. By explicitly challenging technologists to employ anticipation, reflection, inclusiveness and responsiveness when developing new technological objects, such as energy data, the RI paradigm can be used as the foundation and problem shifting can to reveal its multidimensional dynamics.

Box 2.1 Responsible Innovation and Problem Shifting, Side-by-Side and At-a-Glance

There are four classic dimensions of RI (Owen et al. 2013):

- anticipation (i.e., the ability to have foresight in the challenges that technologies may pose and to be able to conduct anticipatory governance that is proactive in nature rather than reactive)
- reflexivity (i.e., actors' and institutions' ability to rethink prevailing conceptions and to contest the apparent inherent amorality and agnostic towards science and innovation's societal implications. It insists researchers to actively interrogate the broader ethical, political, and social dimensions of their work and calls for transparency as a consequence)
- inclusion (i.e., multilevel stakeholders who are involved in all stages of science and technology—from conception to consumption – must be taken into account, and the concept calls for a greater public engagement and more accessible deliberative processes)

• responsiveness (i.e., how systems of innovation can be responsive to the societal and political demands, how to respond to new knowledge and emerging views and perspectives)

These four dimensions have been broadened since then to include other dimensions such as sustainability, care (Burget et al. 2017; Iakovleva et al. 2019), education (van Grunsven et al. 2024)

Problem Shifting happens when efforts to solve a problem in one system, causes problems to emerge:

• in a system at a different time-scale (for instance how industrial agricultural projects designed to decrease hunger cause damage to soil holobionts/quality, which reduce the ecosystem services soil can yield for future generations)

• in a system at a different geographic or spatial scale (for example, when fossil fuel related economic development designed to address poverty increases atmospheric temperatures leading to loss and damage from climate disasters in small island states)

• in a different domain or socio-technical/socio-ecological system (for instance when lock downs designed to prevent healthcare crises lead to educational crises)

Although the emphasis in problem shifting scholarship rests mainly on systems, research suggests that these dynamics can exist between individuals and niches in systems

2.4 The Complementarity of RI and Problem Shifting

The similarity between RI and problem shifting is clear. Both RI and problem shifting advocate for proactively considering the broader social and environmental consequences of technological development. Both call for inclusive and reflexive innovation where marginalized groups are not exploited for the resources that they possess or disproportionately burdened by the unintended consequences of technological advancement. Additionally, both frameworks emphasize the importance of embedding equity, accountability, and sustainability into the design and governance of innovation processes from the outset.

Sharing these fundamental ideas, we argue that problem shifting serves as a valuable complement to RI that describe innovation as a nested or multi-level process. The coupling of RI and problem shifting that we accomplish in this chapter thus provides concrete policy and regulatory guidance, which is an area where RI has frequently been criticized for. At the same time, it provides guidance for observers of problem shifting on how to attribute responsibility in the case of processes, which traverse time, space, and domains. This complementarity is particularly evident given how the concept of problem shifting has been actively applied in the context of international environmental law and policy, resulting in outcomes that are less abstract and more actionable. Instead, RI offers a set of principles which can be applied to evince problem shifting dynamics. Below we discuss in more detail fundamental complementarities between RI and problem shifting:

Underestimates material constraints: RI has been criticized for underestimating the material and infrastructural constraints that shape the feasibility and consequences of innovation. In contrast, the concept of problem shifting is particularly well-suited for identifying such constraints in instances where problems are displaced rather than resolved. For example, when an innovation results in the depletion of water resources or intensifies competition over land use, problem shifting allows us to recognize these outcomes as shifts in the domain of harm—from one resource, region, or community to another. This perspective offers a valuable lens through which we may address the unintended material consequences that RI may overlook.

Social and economic costs of innovation (does not account for tension between innovation): Far from ignoring these dimensions, RI explicitly calls for the integration of diverse perspectives and values to ensure that innovation processes are socially responsive and just. Problem shifting enriches this perspective by providing a more holistic picture by revealing that domain shift of problems. Problem shifting highlights how efforts to address one set of challenges—such as reducing environmental impacts—can inadvertently generate new forms of social or economic harm, often in different domains, disproportionately affecting marginalized communities or future generations. By tracing the displacement of burdens across the three dimensions, namely, spatial, temporal, and domain, problem shifting complements RI's inclusivity goals with a diagnostic tool that makes visible the systemic interdependencies and trade-offs inherent in innovation. It thereby reinforces the need for a more holistic, cross-domain approach to responsibility that goes beyond stakeholder inclusion to account for the broader ripple effects of innovation.

Lacks clear demarcation of responsibilities: RI has been criticized for lacking clear demarcation of responsibilities, a shortcoming that risks enabling forms of organized irresponsibility wherein accountability becomes diffuse or altogether absent. The concept of problem shifting, particularly through its application in international environmental law and governance, offers a potential remedy by introducing legal frameworks that explicitly designate responsible parties. By embedding responsibility within formal institutional and legal structures, problem shifting can enhance the accountability mechanisms that RI seeks but often lacks, thereby reducing the likelihood of responsibility being evaded or obscured in complex innovation systems.

Superficial compliance andstrategic adaptation: All regulatory efforts that aim to moderate or redirect innovation in response to social and environmental challenges are vulnerable to superficial compliance and strategic adaptation by actors seeking to preserve existing trajectories; and RI is no exception. RI arguably may impose a decelerating effect on innovation by demanding complex and often politically sensitive practices, such as anticipatory governance, inclusive stakeholder engagement, and reflexive deliberation. Moreover, through its emphasis on reflexivity and responsiveness, RI introduces the possibility that innovation processes may need to be reversed, revised, or even halted when emerging consequences prove to be socially or ethically untenable. Problem shifting can address this limitation to an extent through its engagement with international law and other national regulatory frameworks. Problem shifting lists action-oriented tasks that international regulatory regimes and organizations and national governments can undertake in order to prevent superficial compliance and strategic adaptation. These strategies can be operationalized through mechanisms such as binding agreements, standardized reporting and monitoring protocols, clearly defined accountability structures, and enforcement provisions that limit the discretionary space for superficial or performative compliance. By embedding responsibility within legal and institutional arrangements, problem shifting offers a structured and enforceable approach to ensuring that innovation that RI was lacking.

Too much emphasis on process than outcomes: RI stresses process of innovation rather than outcomes. However, this is because RI as a theory is attempting to highlight the social and environmental cost of innovation that happen in the process. Therefore, its lack of focus on outcomes is not only intentional but also conceptually grounded in its commitment to reshaping the conditions under which innovation unfolds. By centering attention on the processes, such as anticipation, inclusion, reflexivity, and responsiveness, RI seeks to surface and address the social and environmental costs that are often obscured during the development and deployment of new technologies. Problem shifting also highlights the process rather than outcomes because it identifies the spatial, temporal, and domain shifts of problems that may occur. In doing so, it draws attention to how harms can be displaced or transformed during the course of innovation, reinforcing the need for continuous assessment rather than retrospective evaluation alone. This is not a weakness of the two theories but rather a concrete critique of the prevailing outcome-focused attitude toward innovation, an approach that often overlooks the social, ethical, and environmental dynamics unfolding throughout the innovation process itself.

In short, the problem-shifting framework offers a solution-oriented lens that directly responds to many of the shortcomings identified in critiques of RI. By tracing how challenges migrate and transform across social, technical, and institutional boundaries, problem shifting encourages an approach to governance that is attentive to complexity rather than prescriptive or linear. It invites practitioners to recognize that interventions rarely operate in isolation; they reverberate across domains, producing new pressures even as they alleviate existing ones. This awareness creates space for more adaptive and context-responsive forms of action—strategies that take seriously the specificities of each setting, anticipate unintended consequences, and acknowledge that effective governance cannot be reduced to one-size-fits-all models. The framework thus equips analysts and decision-makers with tools to navigate the evolving landscapes of energy data and innovation more thoughtfully and with greater sensitivity to the interconnected nature of societal challenges.

Conversely, RI is complementary to problem shifting in two ways:

First, one of RI's core strengths lies in its capacity to examine the social conditions under which innovation emerges and to contextualize the societal transformations it produces. RI brings attention to the broader political, cultural, and institutional structures that shape both the direction and impact of innovation. This kind of structural grounding, which is anchored in a deep understanding of social dynamics, is a perspective that problem shifting, with its primarily legal and environmental focus, tends to overlook. In other words, RI extends beyond abstract institutional analysis to include everyday practices, power relations, and value conflicts that shape how technologies are received and used. For example, in the rollout of smart meters, patterns of public trust, perceptions of fairness in billing, and historical relationships between utilities and marginalized communities all influence acceptance and long-term impact. These factors reveal dynamics of consent, participation, and distributive justice that a problem-shifting framework—focused mainly on regulatory design, environmental impact, and legal compliance—may fail to capture. By engaging explicitly with such lived experiences and socio-political contexts, RI provides tools to identify latent tensions, anticipate resistance, and design innovations that are both socially robust and environmentally sound.

Second, one of the least developed dimensions of the problem-shifting framework is its treatment of time, particularly its orientation toward the future. Although the literature offers rich accounts of past events and provides detailed, domain- or place-specific case studies, it pays comparatively little attention to how present decisions create long-term trajectories or to how emerging challenges might unfold. This backward-leaning emphasis limits the framework's capacity to anticipate the cascading effects of technological and environmental change as well as the societal adaptations they may require. In this respect, the concept of anticipatory governance advanced within the RI tradition offers a valuable complement. Anticipatory governance calls for systematic foresight, scenario building, and the inclusion of diverse stakeholders to envision alternative futures and to test policies against a range of plausible developments. Integrating these practices into problem shifting analysis would strengthen its ability to devise forward-looking strategies, identify latent risks before they crystallize, and embed flexibility into policy design. Such an extension would not only enhance the temporal depth of problem shifting but also better equip it to address the long-term uncertainties inherent in complex socio-technical systems.

These two discussion points illustrate how RI enriches the problem-shifting framework by supplying forms of insight that are otherwise absent or underdeveloped. RI's attention to the social elements of innovation (i.e., the lived experiences, value disputes, and power relations that shape how technologies are taken up) adds a layer of analytic depth that complements problem shifting's focus on regulatory structures and environmental consequences. At the same time, RI's orientation toward anticipation and future-making offers a means of counterbalancing the retrospective tendencies of problem shifting, enabling a more expansive engagement with long-term trajectories and uncertainties. When these perspectives are brought into dialogue, they form a more comprehensive approach to understanding and governing socio-technical change.

2.5 Where Energy Data Fits in

We argue that RI and problem shifting offer complementary and valuable frameworks for discussing energy data. RI provides a normative and holistic orientation that situate energy data within broader societal and ethical contexts. In contrast, problem shifting draws attention to the regulatory and legal mechanisms necessary to operationalize these goals, offering concrete tools to identify and address the displacement of environmental and social harms. Together, these frameworks enable both critical reflection and actionable guidance in the governance of energy data. Moreover, the combined lenses of RI and problem shifting enable energy data to be understood as the technological artifact that it is, rather than as an abstract and disembodied entity that is inherently elusive or immaterial.

Energy data can be understood as a form of problem shifting. Initially collected to enhance grid stability, enable effective energy management, and ensure a steady and reliable supply of electricity, it was also heralded as a critical enabler for the integration of renewable energy sources. However, while advancing these goals, the proliferation of energy data has given rise to significant privacy concerns (Lee and Hess 2021) and new sustainability challenges related to data storage and data centers (Lee et al. 2024). Moreover, energy data exemplifies problem shifting in the regulatory domain, as efforts to promote transparency have encountered resistance. In many countries, access to energy data is tightly controlled by governments, limiting its utility while concentrating authority over its use even for research. This underscores how energy data, although positioned as a solution to existing energy system challenges, simultaneously generates new forms of social, environmental, and institutional complexity.

Energy data is deeply embedded within social and political contexts, as patterns of electricity use are shaped by factors such as climate, daily routines, cultural practices, economic conditions, and personal values (Fong et al. 2007; Sanquist et al. 2012). These factors are situated within broader societal structures that govern energy infrastructures, regulatory frameworks, and pricing mechanisms. As such, energy data is not a neutral or purely technical artifact, but instead reflects and reinforces existing social arrangements and power relations. RI offers a framework for critically examining how data is produced, accessed, and used, and for ensuring that these processes align with principles of inclusion, accountability, and social justice. By foregrounding the social dimensions of technological systems, RI provides the tools to interrogate whose interests are served by energy data and how its governance can be made more equitable and responsive.

In this sense, by bringing together the insights of RI and problem shifting—and by demonstrating the ways in which these frameworks complement and enrich one another—this book seeks to move beyond the identification of social, political, and environmental problems associated with energy data. Our goal is not merely to diagnose shortcomings or catalogue risks, but to articulate pathways toward more just, accountable, and sustainable forms of data governance. For example, in earlier work (2025) we demonstrated that individuals are often willing to share highly sensitive vehicle data including detailed location histories, driving performance metrics, and in some cases, in car-video footage, when they perceive that such disclosure advances broader social goods such as public safety and environmental sustainability. Their findings illustrate that data governance cannot be grounded solely in abstract notions of privacy but should also account for the values, trade-offs, and collective aspirations that shape real world decision making. This type of insight underscores the value of a solution focused approach in which data practices are not considered to be risks to be avoided, but how governance and collective aspirations can be designed to enable socially meaningful uses of data while still safeguarding rights and limiting potential harms. In doing so, the book aspires to contribute to a more affirmative conversation about energy data by acknowledging existing challenges while equipping policymakers, practitioners, and scholars with tools to reshape data practices in ways that better serve both present and future generations.

2.6 Conclusion

Energy data takes on different meanings and implications at each stage of its lifecycle, and its governance cannot be fully understood without the social, political, and institutional landscapes in which it is embedded. Bringing RI and problem shifting into conversation allows us to move beyond narrowly technical accounts of data systems and meter focused data management and to approach energy data as a sociotechnical artifact: one whose functions, risks, and benefits emerge through the interplay of technological design, regulatory choices, and societal values.

The chapters that follow draw on this integrated framework to explore how energy data is produced, shared, stored, and governed, and how these practices generate both opportunities and vulnerabilities. RI and problem shifting together illuminate the ways in which energy data, which was initially justified as a means of improving grid resilience and supporting renewable integration, has simultaneously introduced new privacy risks, amplified existing inequities, and contributed to the growing environmental pressures of digital infrastructures.

Understanding these dynamics is essential for developing governance models that do not take energy data at face value but rather confront the complex assemblage of relationships and consequences it entails. Energy data is shaped by climatic conditions, household routines, cultural norms, economic circumstances, and institutional arrangements; it reflects longstanding social patterns even as it introduces new forms of oversight and control. By placing these elements at the center of analysis, RI helps us to examine whose interests are prioritized in energy data practices and how these choices influence inclusion, accountability, and equity.

At the same time, recognizing energy data as a form of problem shifting makes visible the unintended consequences that accompany its expansion. Sustainability impacts, opaque data ownership structures, restrictions on access, which are often justified in the name of reliability or innovation, reveal that energy data systems can reproduce the very problems that they claim to solve. Identifying and addressing these shifts requires governance tools attuned to both current challenges and emerging risks.

This chapter lays the theoretical groundwork for a more deliberate and reflective approach to energy data governance – one that resists the pure technical fixes and one that instead embraces a fuller consideration and social context, environmental impact, and institutional responsibility. This combined perspective of RI and problem shifting not only sharpens our critique but also clarifies the kinds of interventions needed to shape more just and sustainable energy data futures.

References

Alami H, Lehoux P, Shaw SE, Niang M, Malas K, Fortin J-P (2024) To what extent can digital health technologies comply with the principles of responsible innovation? Practice-and policy-oriented research insights regarding an organisational and systemic issue. Int J Health Policy Manag 13:8061. https://doi.org/10.34172/ijhpm.8061

Amutha C, Gopan A, Pushbalatatha I, Ragavi M, Reneese JA (2024) Nanotechnology and governance: regulatory framework for responsible innovation. In: George SC, Tawiah B (eds) Nanotechnology in societal development. Singapore: Springer Nature, pp 481–503

Beck U (1992) Risk society: towards a new modernity. Sage Publications, London

Beck U (1997) Subpolitics: ecology and the disintegration of institutional power. Organ Environ 10(1):52–65. https://doi.org/10.1177/0921810697101008

van den Bergh J, Folke C, Polasky S, Scheffer M, Steffen W (2015) What if solar energy becomes really cheap? A thought experiment on environmental problem shifting. Curr Opin Environ Sustain 14:170–179. https://doi.org/10.1016/j.cosust.2015.05.007

Blok V, Lemmens P (2015) The emerging concept of responsible innovation. Three reasons why it is questionable and calls for a radical transformation of the concept of innovation. Cham: Springer International Publishing

Boateng FG, Klopp JM (2024) The electric vehicle transition: a blessing or a curse for improving extractive industries and mineral supply chains? Energy Res Soc Sci 113:103541. https://doi.org/10.1016/j.erss.2024.103541

Buhmann A, Fieseler C (2021) Towards a deliberative framework for responsible innovation in artificial intelligence. Technol Soc 64:101475. https://doi.org/10.1016/j.techsoc.2020.101475

Burget M, Bardone E, Pedaste M (2017) Definitions and conceptual dimensions of responsible research and innovation: a literature review. Sci Eng Ethics 23(1):1–19. https://doi.org/10.1007/s11948-016-9782-1

Cohen T, Stilgoe J, Stares S, Akyelken N, Cavoli C, Day J, Dickinson J, Fors V, Hopkins D, Lyons G, Marres N, Newman J, Reardon L, Sipe N, Tennant C, Wadud Z, Wigley E (2020) A constructive role for social science in the development of automated vehicles. Transp Res Interdiscip Perspect 6:100133. https://doi.org/10.1016/j.trip.2020.100133

Cummings CL, Kuzma J, Kokotovich A, Glas D, Grieger K (2021) Barriers to responsible innovation of nanotechnology applications in food and agriculture: a study of US experts and developers. NanoImpact 23:100326. https://doi.org/10.1016/j.impact.2021.100326

Hoop D, Evelien AP, Romijn H (2016) Limits to responsible innovation. J Responsible Innov 3(2):110–134. https://doi.org/10.1080/23299460.2016.1231396

Doostizadeh M, Ghasemi H (2012) A day-ahead electricity pricing model based on smart metering and demand-side management. Energy 46(1):221–230. https://doi.org/10.1016/j.energy.2012.08.029

Egbue O (2012) Assessment of social impacts of lithium for electric vehicle batteries. In: IISE annual conference. Proceedings, pp 1–7. https://www.proquest.com/docview/1151086376/abstract/FEE82BA1A6754ACFPQ/1

Eyre HA, Ellsworth W, Ernestine F, Manji HK, Berk M (2020) Responsible innovation in technology for mental health care. Lancet Psychiatry 7(9):728–730. https://doi.org/10.1016/S2215-0366(20)30192-9

Fong W-K, Matsumoto H, Lun Y-F, Kimura R (2007) Influences of indirect lifestyle aspects and climate on household energy consumption. J Asian Arch Build Eng 6(2):395–402. https://doi.org/10.3130/jaabe.6.395

Galantino MG (2022) Organised irresponsibility in the post-truth era: beck's legacy in today's world at risk. Ital Sociol Rev 12(8S):971–971. https://doi.org/10.13136/isr.v12i8S.598

Gao Y, Skutsch M, Masera O, Pacheco P (2011) A global analysis of deforestation due to biofuel development. Cent Int for Res (CIFOR). https://doi.org/10.17528/cifor/003506

Geels FW (2014) Regime resistance against low-carbon transitions: introducing politics and power into the multi-level perspective. Theory Cult Soc 31(5):21–40. https://doi.org/10.1177/0263276414531627

Geels FW (2019) Socio-technical transitions to sustainability: a review of criticisms and elaborations of the multi-level perspective. Curr Opin Environ Sustain 39:187–201. https://doi.org/10.1016/j.cosust.2019.06.009

Geels FW, Sovacool BK, Schwanen T, Sorrell S (2017) The socio-technical dynamics of low-carbon transitions. Joule 1(3):463–479. https://doi.org/10.1016/j.joule.2017.09.018

Genus A, Iskandarova M (2018) Responsible innovation: its institutionalisation and a critique. Technol Forecast Soc Chang 128:1–9. https://doi.org/10.1016/j.techfore.2017.09.029

Genus A, Stirling A (2018) Collingridge and the dilemma of control: towards responsible and accountable innovation. Res Policy 47(1):61–69. https://doi.org/10.1016/j.respol.2017.09.012

German L, Pacheco P, Schoneveld G (2011) The social and environmental impacts of biofuel feedstock cultivation: evidence from multi-site research in the forest frontier. Ecol Soc 16(3):24. https://doi.org/10.5751/ES-04309-160324

van Grunsven J, Stone T, Marin L (2024) Fostering responsible anticipation in engineering ethics education: how a multi-disciplinary enrichment of the responsible innovation framework can help. Eur J Eng Educ 49(2):283–298. https://doi.org/10.1080/03043797.2023.2218275

Gumbert T (2022a) Envrionmental governance and the organization of irresponsibility. In: Gumbert T (ed) Responsibility in environmental governance: unwrapping the global food waste dilemma. Cham: Springer International Publishing, pp 23–50

Hatch ME, Graff M (2024) Housing costs are not a monolith: the association between neighborhood energy burdens and eviction filing rates. Cities 150:104995. https://doi.org/10.1016/j.cities.2024.104995

Herrmann H (2023) What's next for responsible artificial intelligence: a way forward through responsible innovation. Heliyon 9(3). https://doi.org/10.1016/j.heliyon.2023.e14379

Hess DJ (2024) Microgrids and the politics of sustainability transitions: a sociotechnical, multi-coalition perspective. Environ Innov Soc Trans 51:100839. https://doi.org/10.1016/j.eist.2024.100839

van den Hoven J (2022) Responsibility and Innovation. J Responsible Innov 9(1):133–137. https://doi.org/10.1080/23299460.2022.2050570

Iakovleva T, Ødegard A, Willumsen E (2019) Towards the responsible innovation outcomes: role of anticipation and inclusiveness—ProQuest. In: ISPIM conference proceedings. https://www.proquest.com/docview/2297091757?pq-origsite=gscholar&fromopenview=true&sourcetype=Conference%20Papers%20&%20Proceedings.

Iakovleva T, Oftedal EM, Bessant J (2019) Chapter 1: responsible innovation in digital health

Kim R, van Asselt H (2016) Chapter 23: global governance: problem shifting in the anthropocene and the limits of international law in: research handbook on international law and natural resources. P. c 584 in Research handbook on international law and natural resources. Elgar

Kokotovich AE, Kuzma J, Cummings CL, Grieger K (2021) Responsible innovation definitions, practices, and motivations from nanotechnology researchers in food and agriculture. NanoEthics 15(3):229–243. https://doi.org/10.1007/s11569-021-00404-9

Lee D, Hess DJ (2021) Data privacy and residential smart meters: comparative analysis and harmonization potential. Util Policy 70:101188

Lee D, Hess DJ, Heldeweg MA (2022) Safety and privacy regulations for unmanned aerial vehicles: a multiple comparative analysis. Technol Soc 71:102079. https://doi.org/10.1016/j.techsoc.2022.102079

Lee D, Long LAN, Jansma SR (2024) Data donation: using the gift relationship framework to address privacy and environmental issues of energy consumption data collection. Energy Res Soc Sci 114:103596. https://doi.org/10.1016/j.erss.2024.103596

Lee D, Long LAN (2025) Privacy challenges of automated vehicles: merging contextual integrity and responsible innovation frameworks. Transp Res Interdiscip Perspect 32:101536. https://doi.org/10.1016/j.trip.2025.101536

Macnaghten P (2020) The making of responsible innovation. Cambridge University Press, Cambridge United Kingdom

Mahler V, Girard R, Kariniotakis G (2022) Data-driven structural modeling of electricity price dynamics. Energy Econ 107:105811. https://doi.org/10.1016/j.eneco.2022.105811

Martinez N (2022) Theorizing the devices of sociotechnical governance: systemic practices, visions, and dynamics of change in Mexico's energy transition. Energy Res Soc Sci 90:102626. https://doi.org/10.1016/j.erss.2022.102626

Naughton B, Dopson S, Iakovleva T (2023) Responsible impact and the reinforcement of responsible innovation in the public sector ecosystem: cases of digital health innovation. J Responsible Innov 10(1):2211870. https://doi.org/10.1080/23299460.2023.2211870

Owen R, Pansera M (2019) Chapter 2: responsible innovation and responsible research and innovation. In: Handbook on science and public policy. Elgar, pp 26–48

Owen R, Stilgoe J, Macnaghten P, Gorman M, Fisher E, Guston D (2013) A framework for responsible innovation. In: Responsible innovation: managing the responsible emergence of science and innovation in society. West Sussex, United Kingdom, Wiley, pp 27–50

Pandza K, Ellwood P (2013) Strategic and ethical foundations for responsible innovation. Res Policy 42(5):1112–1125. https://doi.org/10.1016/j.respol.2013.02.007

Pavie X, Scholten V, Carthy D (2014) Responsible innovation: from concept to practice. World Scientific Publishing Co., Ltd., Singapore

van de Poel I, Sand M (2021) Varieties of responsibility: two problems of responsible innovation. Synthese 198(19):4769–4787. https://doi.org/10.1007/s11229-018-01951-7

Raven R (2024) Strategic niche management: past, present, and future

Sanquist TF, Orr H, Shui B, Bittner AC (2012) Lifestyle factors in U.S. residential electricity consumption. Energy Policy 42:354–364. https://doi.org/10.1016/j.enpol.2011.11.092

Stahl BC (2022) Responsible innovation ecosystems: ethical implications of the application of the ecosystem concept to artificial intelligence. Int J Inf Manag 62:102441. https://doi.org/10.1016/j.ijinfomgt.2021.102441

Stilgoe J (2013) Front matter: why responsible innovation. In: Responsible innovation: managing the responsible emergence of science and innovation in society. West Sussex, United Kingdom, Wiley pp 1–24

Stilgoe J (2021) How can we know a self-driving car is safe? Ethics Inf Technol 23(4):635–647. https://doi.org/10.1007/s10676-021-09602-1

Stilgoe J, Mladenović M (2022) The politics of autonomous vehicles. Humlties Soc Sci Commun 9(1):1–6. https://doi.org/10.1057/s41599-022-01463-3

Stilgoe J, Owen R, Macnaghten P (2013) Developing a framework for responsible innovation. Res Policy 42(9):1568–1580. https://doi.org/10.1016/j.respol.2013.05.008

Wiek A, Foley RW, Guston DH, Bernstein MJ (2016) Broken promises and breaking ground for responsible innovation—intervention research to transform business-as-usual in nanotechnology innovation. Technol Anal Strat Manag 28(6):639–650. https://doi.org/10.1080/09537325.2015.1129399

Wood Hansen O, Van Den Bergh J (2023). In: Amon C (ed) Environmental problem shifting from climate change mitigation: a mapping review. PNAS Nexus 3(1):ad448. https://doi.org/10.1093/pnasnexus/pgad448

Yang Y, Bae J, Kim J, Suh S (2012) Replacing gasoline with corn ethanol results in significant environmental problem-shifting. Environ Sci Technol 46(7):3671–3678. https://doi.org/10.1021/es203641p

Ye X, Zhang Z, Qiu Y (2023) Review of application of high frequency smart meter data in energy economics and policy research. Front Sustain Energy Policy 2. https://doi.org/10.3389/fsuep.2023.1171093

Chapter 3
Governing Energy Data

3.1 Intersections of Privacy, Justice, and Sustainability

Energy data belongs to a complex of technologies, processes, policies, and practices that together make up the (renewable) energy transition (Dekker and Van Est 2020). The focus of this chapter is on policies. Although there are many diverse definitions of policy (Cairney 2012), this chapter adopts a broad definition of policy as "a formal decision or plan of action adopted by an actor, be it an individual, organization, business, government etc., in order to achieve a particular goal." (Richards and Smith 2008, p. 1). The goal that we focus on is energy data, and we interrogate the aims that may underlie this goal (e.g., sustainability, efficiency, affordability, security) and the principles that do or do not guide its production, storage, and distribution (e.g., privacy, equity, sustainability). Energy data and its societal impact is also shaped then by the goals set in the language of policies, and the actions taken to achieve those ends.

One thing that is implicit in all definitions of policy is the idea that policies are a means by which societies support or slow change processes (Borrás and Edquist 2013; Flanagan et al. 2011; Howlett et al. 2015). When it comes to energy systems we depart from the understanding of energy systems as socio-technical systems and are therefore predisposed to stability rather than change. Underpinning this stability is a web of technical and social elements that resist change (Latour 2005). Policy is a key social element in this configuration (Schaefer et al. 2021). Continued strong alignment between different social groups that is held together by a complex system of rules make only minimal changes possible within the energy system (Geels 2002; Geels et al. 2017; Nguyen Long et al. 2024). Innovation policies are seen as a requirement to protect game changing technologies—like energy data—in niches embedded in what is otherwise a harsh environment for innovation (Geels 2002).

One way to protect such a technology and its development is to make it "safe for use" by developing regulatory guardrails that can reassure the public about the safety of its use. However, this task is neither simple nor is it straightforward in

L. A. Nguyen Long and D. Lee, *Energy Data*, Lecture Notes in Energy 46,
https://doi.org/10.1007/978-3-032-19171-7_3

the case of energy data. Energy data promotes innovation, and innovation and its governance are often subservient to a range of other social goals (e.g., growth, the environment or the health system) that often lie beyond policymaking discussions around the innovation itself (e.g., Borrás and Edquist 2013; Branscomb and Florida 1998; Smith 1994). Moreover, energy data policies can vary across jurisdictions, which can complicate its use(s). For example, the differences in national energy data regulations significantly hinders energy data sharing between European states (Kashyap et al. 2025). The fragmented and complex nature of energy data governance has implications for other policy subsystems.[1]

Policies do not always perform as intended. For instance, they may simultaneously trigger different outcomes, some intended others not. A policy may inadvertently give rise to a perverse incentive, which unintentionally rewards behaviors directly opposed to the policy's original purpose and leads to negative or counterproductive outcomes. This has been seen, for example, in the U.S. energy market, where energy subsidies and regulated low prices have in the past unintentionally intensified energy consumption and produced energy waste (Gill 1976). As the example suggests, energy policies can fail and the literature is chock full of examples of when this has come to pass in different regulatory environments and social contexts (Ardoin and Grady 2006; Rosenow and Eyre 2016; Yoon and Sim 2015). This can also be the case for the policy instruments that regulate and enable energy data. Given these and other complications, the jury is still out on whether energy data can help countries and communities meet their energy transition ambitions while upholding principles of energy justice.

3.2 Energy Data Policy Mixes

As sociotechnical systems change to meet the unpredictable demands and outcomes of sustainability transitions, there is a continuing need to realign policies with the rapidly evolving conditions on the ground (Hoppmann 2013; Loorbach and Rotmans 2010). When existing institutional arrangements are out of sync with the policy instruments that are introduced to govern the technologies driving the transition, conflicts between participants, programs, and goals can arise (Edmondson et al. 2019; Patashnik and Zelizer 2009). In such cases, the likelihood of problem shifting is expected to be particularly high.

To review, problem shifting occurs when a system degrades as an (un)-intended consequence of efforts to improve the performance of another (Kim and Van Asselt 2016). Problem shifting can be temporal (problems deferred to future generations), spatial (problems shifted to other places), domain driven (problems shifted to other domains), or a combination of any number of these dimensions. Problem shifting may manifest—this chapter argues—as compromises with regards to one social goal,

[1] Policy subsystems are networks of actors, institutions, and processes that focus on a specific policy issue (e.g. elderly care) or set of related issues (e.g. energy transitions) over time.

e.g., privacy, in service of another, e.g., sustainability. This lens can be used to reveal how interdependencies lead to tradeoffs between social goals that ultimately benefit energy companies and private intermediaries at cost to energy consumers and other constituencies. From an energy justice lens, the emergence of harms in the course of implementing energy data policy can be considered policy failure (Sokołowski and Heffron 2022) (Fig. 3.1).

Compounding and amplifying problem shifting dynamics triggered by policy misalignment is the tendency of policies with visible successes (Kingdon 2003; Makse and Volden 2011) and imperceptible consequences to become mobile (Lovell et al. 2023). This is especially in situations of so-called 'fast policy' wherein the international circulation of policy examples and ideas is hastened (Peck and Theodore 2015). The diffusion of energy policy is a well-studied phenomenon. Although there are many studies on the performance of energy policy instruments (e.g., Kim 2020), the patterns that have been observed in energy policymaking suggests that performance is not the main factor determining diffusion, which align with what has been seen across policy domains (Karch et al. 2016; Shipan and Volden 2008). Instead of performance, evidence suggests that national decisionmakers are more likely to copy energy policies of countries that they share similarities in infrastructure and

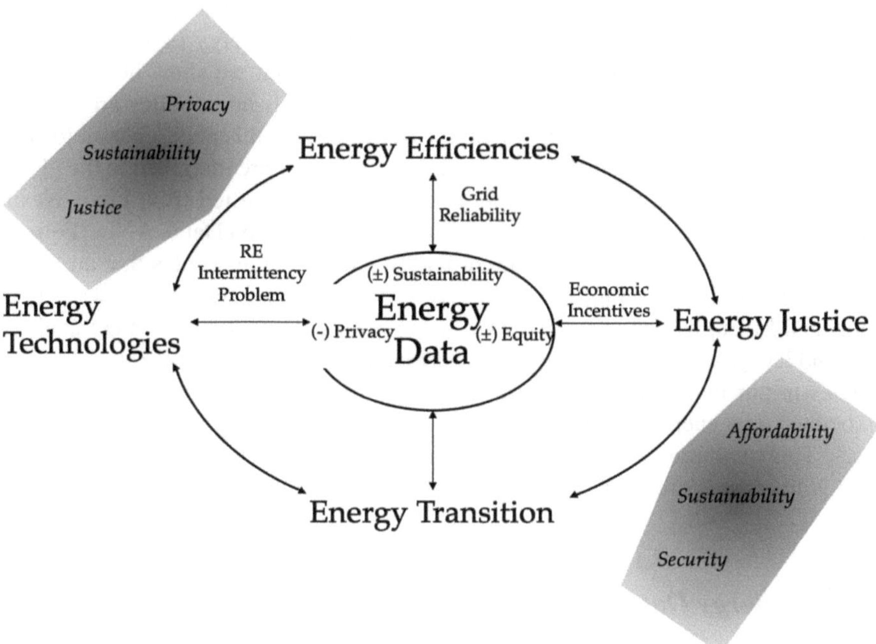

Fig. 3.1 Energy data challenges and opportunities. *Caption: The complex of energy data interdependencies gives rise to numerous opportunities related to the energy transition, e.g. energy affordability, sustainability, and security, while also creating problems and harms in other domains related to privacy, sustainability, and socio-economic injustices*

electricity-use patterns with (Baldwin et al. 2019; Nicholson-Crotty and Carley 2016). In developing countries, energy policy adoption is also highly influenced by donor preferences (Baldwin et al. 2019; Cia Alves et al. 2019), which increases the likelihood that an adopted policy will not match the needs on the ground.

Policy mixes emphasize dynamic interactions between policy instruments to offer a possible mechanistic explanation for the problem shifting phenomenon. Changes in one instrument or strategy of the policy mix can, for example, have an impact on institutions and on resource use and allocation. This can carry consequences for or stimulate responses from stakeholders of other policy domains affected by the mix (Patashnik and Zelizer 2009). But perhaps a more insidious and influential consequence of changes in the mix is its interpretative consequences (Pierson 1996), that change how stakeholder understand and anticipate policy outcomes (Edmondson et al. 2019). Policies themselves are often interpreted and reinterpreted in the light of changing rationales (Flanagan et al. 2011).

3.3 The Energy Data Policy Mix

In this section, we contemplate the policy mixes, games, and problem shifting dynamics in energy data. To that end, we consider how energy data as a sustainability strategy justifies the use of smart grids. Thereafter, we show how this may shift problems from the energy policy arena to the digitalization policy arena in the form of privacy threats and social inequity. To interrogate this phenomenon further, we then look at how these different energy data outcomes are imagined in the global mainstream media by using a dataset of news articles harvested from Lexis-Nexus on 16 October 2025 with the keywords "Energy Consumption Data" AND "Policy." Table 3.2 (in the Appendix) lists and provides details on the articles that are in the text corpus. As can be seen from Table 3.1, the text corpus includes 39 news articles published in English between 2010 and 2024. Using a qualitative content analysis (Elo and Kyngäs 2008; Kyngäs 2020), we critically synthesize the content of the news articles in our text corpus to identify trends, for example—the dominant concepts associated with energy data policy in order to identify recurring patterns across these cases. This qualitative content analysis is applied to map the different concepts that arise with regards to energy data policy in the print media discourse.

3.3.1 Energy Data Policies? What Might the Mix Look Like?

A mix encompasses a range of policy instruments. Policy instruments can be used to encourage behaviors (e.g., offering households incentives like financial subsidies for installing solar panels or installation) or discourage behaviors (e.g., punitive actions like fines for companies whose emissions than their allotted CO_2 emission allowances) (Schneider and Ingram 1990). Instrument choice can hang on several

Table 3.1 Energy data policy instruments

Instrument	General description	Examples
Regulatory instruments	Rules, Laws, and Directives that use hierarchical and legal power to oblige policy targets to comply with the mandated actions	Making the installation of smart meters compulsory for households and businesses
Economic and financial instruments	Are taxes, subsidies, inducement, grants, fines, etc. that encourage or discourage behaviors by lowering or raising adoption barriers	Rebates on "smart" appliances
Soft instruments	Include campaigns, codes of conduct, recommendations, voluntary agreements that leverage that rely on norms and moral dispositions to encourage or discourage actions on the part of policy targets	Information sharing platforms where energy consumers can compare smart meters and appliances

factors, including their expected effects over time, their cost-benefit profiles, and the kinds of conditions under which they are likely to succeed (McDonnell and Elmore 1987). Three instrument types have been associated with innovation policies, they are regulatory, financial, and soft (Borrás and Edquist 2013), and in the case of energy data policies, each of these are reflected in the media reports that were analyzed (See Table 3.1).

Regulatory tools expect compliance from those being targeted by policies. They oblige policy targets to comply with the rules, laws, and directives issued from the top-down, (Borrás and Edquist 2013; McDonnell and Elmore 1987; Schneider and Ingram 1990) which are enforced via legitimate authorities of the state (Capano and Howlett 2020). Examples of regulatory tools used to govern energy data that we found in the media reports are geared mainly towards enabling the collection of energy consumption data from households. This includes reports on laws that make it compulsory for households to install smart meters ~Item #38, Item #17, and Item #29.

The newspaper reporting indicated that some governments were leaning towards adopting such mandates (e.g., U.K., Sri Lanka) regardless of disputes from citizen interest groups. For instance, in the U.K., one report observes that, "*The Parliamentary Select Committee sensed public opposition to smart meters, but the Government's latest consultation says it intends to make them compulsory - going back on promises. All energy suppliers will have to collect your energy consumption data. Promises that your data would never be commercialised have evaporated before. Data is valuable*"~Item #38. Other countries (e.g., Germany) chose to abstain from adopting such regulations because of those concerns and because of research suggesting that customer savings from the installation of these devices may not materialize.

Financial instruments use incentives to encourage or discourage behaviors (e.g., rewards or punishments) (Borrás and Edquist 2013; McDonnell and Elmore 1987). They may also enable certain actions to lower adoption barriers, for instance by

Table 3.2 Media articles included in the text corpus

Item#	Title	Author	Published in	Date published
1	Action plan to minimize electricity consumption		IPR	2014-08-14
2	U-M, community partners tackle energy insecurity in three Detroit neighborhoods	Jim Erickson	University Wire	2020-11-23 Pg. 1
3	Report says Maryland emissions growing	Megan Hartley	The Capital (Annapolis, MD)	2007-04-13 Pg. A4
4	Meter readers face threat of dogs		Warwick Daily News (Queensland)	Pg. 3 2019-01-12
5	Can you getting a smart meter really stop Britain relying on imported gas?	Jeff Prestridge & Rachel Rickard Straus	MailOnline	2023-10-31
6	Breathing space on climate policy	Les Coleman	The Australian	2009-12-08, Pg. 12
7	China's carbon emissions from fossil fuels may be 14% lower than thought New analysis takes fuel quality into account for first time, but China is still world's largest emitter of greenhouse gasses	Emma Howard	The Guardian	2015-08-19
8	Power prices online: new tool to save costs	Matt Johnston	Herald Sun/Sunday Herald Sun (Melbourne, Australia)	2013-04-02 Pg. 11
9	Sault's climate change coordinator works on mitigation; project includes development of policies and actions to reduce GHG emissions		Sault Ste. Marie This Week	2019-09-05 Pg. A1
10	Agencies rush to install smart meters	Tim Kauffman	Federal Times	2009-08-10; Pg. 11
11	The complex yet rewarding path to greening the built environment society-wide buy-in and mobilisation of resources across multiple stakeholder groups needed but investments will pay back in time	Michelle Quah	The Business Times Singapore	2022-02-28
12	Campbell's hidden $200-million tax cut	Mark Jaccard	The Vancouver Sun (British Columbia)	2011-04-12; Pg. A11

(continued)

Table 3.2 (continued)

Item#	Title	Author	Published in	Date published
13	India as the new rising star		CE Noticias Financieras	2024-03-15
14	Pulse energy aloft on energy-saving software offering; Vancouver company finds local and international markets receptive to its ideas	Scott Simpson	The Vancouver Sun	2011-02-24; Pg. D3
15	E-Quest helping green up Beijing building project	Lee Batdorff	Crain's Cleveland Business	2002-11-18; Pg. 29
16	States urged to fight global warming		St. Louis Post-Dispatch (Missouri)	1990-07-30; Pg. 5I
17	Energy management where are we?		Daily News (Sri Lanka)	2010-05-07
18	Utility companies start hawking appliances US utilities are helping to sell energy efficient gear in an attempt to give away more rebates to customers. Will the strategy work—and will buying new stuff really save energy?	Ucilia Wang	The Guardian	2015-05-13
19	Never mind the new energy meters. It's other "smart" devices at home that should really frighten you	Alex Brummer	The Daily Mail (London)	2018-01-30
20	"The public sector must experiment with artificial intelligence to be able to decide its limits"	José Luis Escrivá	CE Noticias Financieras	2024-02-29
21	The business case for energy efficiency Decarbonisation is not only good for the environment, it's good for the bottom line	Mark Hilliard	The Irish Times	2015-11-27; Pg. 6
22	Green schools	Jacqueline Yallop	The Times Educational Supplement	2004-10-29; Pg.11
23	Starting the year on a green slate	Yvonne Yoong	New Straits Times (Malaysia)	2013-01-04; Pg. 6
24	Cambridge plugs into current take on energy use and conservation	Kevin Swayze	Waterloo Region Record	2009-05-28; Pg. B1
25	Are we electricity customers or hostages?		Isle of Man Today	2015-08-06

(continued)

Table 3.2 (continued)

Item#	Title	Author	Published in	Date published
26	Smart thinking fuels Vecto's foray into Britain		The New Zealand Herald	2012-08-23
27	Want real insight? Let products do the talking: from the store to the home, marketers are mining surprising new sources for data	Kate Kaye		2013-05-20; Pg. 14
28	Juice energy conservation meter's running consumer electronics: displays that reveal how much electricity your home is using can give you a nasty—but informative—surprise		Star Tribune (Minneapolis MN)	2008-03-14; Pg. 6D
29	States weighing pros, cons of smart meters	Susan Haigh and Stephen Singer	Pantagraph (Bloomington, Illinois)	2011-11-14; Pg. A1
30	Overheating meters probed after Peco halts installation	Andrew Maykuth	The Philadelphia Inquirer	2012-08-28
31	The shift from voice to data communications	Emma Okonji	This Day (Lagos)	2015-01-15
32	South Elgin, ComEd agree to water meter deal		Chicago Daily Herald	2021-06-23; Pg. 0
33	Regulators investigating reports of seriously overheating electric meters	Andrew Maykuth	The Philadelphia Inquirer	2012-08-29; Pg. A01
34	Benefit of smart meters in doubt; Tiny savings after huge cash outlay	Darcy Henton	Edmonton Journal (Alberta)	2010-05-22; Pg. B9
35	Manchester can lead the way in research and development	Juergen Maier	Manchester Evening News	2013-06-20; Pg. 8
36	Centre invites smart-grid project proposals		India Business Insight Database	2015-08-13; Pg. 20
37	Building greener and smarter How can smarter office buildings better serve the needs of businesses and employees?		The Business Times Singapore	2012-10-22
38	No Headline In Original		Essex Chronicle	2015-04-03, Pg. 18
39	Privacy commissioner to Investigate smart meters		Alberni Valley Times (British Columbia)	2011-07-29; Pg. A5

building policy targets capacity for action (Schneider and Ingram 1990). When it comes to energy data, the newspapers report on the use of rebates to encourage households to install smart meters and appliances with sensors in the U.S. These rebates are considered necessary because, "*the lure of savings alone apparently isn't enough to propel many customers through the hassle of replacing their appliances*"~Item #15. Another article reports on how federal government agencies (e.g., Veteran's Affairs, Department of Commerce) in the U.S. are applying stimulus funds received from the central government to install smart meters and comply with the 2005 Policy Act before the deadline~Item #10.

Instead, soft policy instruments (e.g., campaigns, codes of conduct, recommendations, voluntary agreements) rely on voluntary compliance with policy recommendations (Borrás and Edquist 2013). They expect policy targets to comply with policy because it coheres with their moral or value preferences. These "calls to conscience" may help policy targets learn more about an issue—e.g., through a public awareness campaigns—and to act in accordance with that learning (Schneider and Ingram 1990). Alternative, symbolic tools use gestures and public persuasion to mobilize public action (Schneider and Ingram 1990). There are only a few examples of soft instruments in the newspapers. One article describes an information sharing platform used by state governments to encourage household adoption of smart meters ~Item #8, for instance Connecticut state government's platform that: "*ranks various appliances' energy efficiency and pricing, that allows consumers to comparison shop and claim rebates online*"~Item #15.

As might be expected under a policy-mix scenario, some articles call for hybrid approaches. For example, an article that describes a combination of soft and financial tools. For example, they used framings such as "*We are at a point where what we don't have is going out of fashion and what we do have is coming into fashion*" and "*Paris is going to be, in my view, a much stronger sense that things are going one way and we are getting out of fossil fuels*" while also insisting that, "*We need policy to evolve in terms of [energy] price and incentives and disincentives so we can start to change*"~Item #21.

However, beyond regulating energy data, energy data itself can be used to achieve other policy goals such as sustainability and energy affordability. It may also become a target of policy to the extent that it violates other policy or social goals, as may also be seen with sustainability, equity, and privacy. We discuss these points in more detail in what follows.

3.3.2 Energy Efficiency in the Energy Data Policy Mix

One of the main aims for the use of energy data technologies like appliances with sensors, smart meters, and smart grids is to achieve energy efficiency. The ability to generate real time energy data is described as helping power distribution officers the capacity to, "*balance power use while giving homeowners and businesses choices of when they want to buy power at what cost*"~Item #24. This allows energy systems to be more reliable and agile and more effectively manage fluctuations in energy use~Item #29. Energy data and the resulting efficiency is also tightly connected to energy affordability in the media discourse. This benefit of energy technologies and the energy data they produce is a dominant idea in media reporting on energy transitions and green energy discourses~Item #14, Item # 26, Item #27, Item #11. For instance, in an article on digitalization, generally, smart grids are lauded as being able to "*allow homeowners to better regulate their own energy consumption and costs*"~Item #27.

However, there are sectors and geographies that are not carried away by the enthusiasm over smart meters. For instance, rural electrification associations and consumer groups in Alberta called for more oversight and careful review of the smart meter rollout. They furthermore argue that there may not be a strong need for the technology in their region, "*Alberta doesn't have the peak load problem other jurisdictions have because its large industrial users already have time-of-use metering that enables them to shift their load to off-peak hours when necessary*" These stakeholders commissioned a paper that concluded "*The requirement to provide real-time, two-way communications with rural meter sites, to capture hourly consumption and provide for automated interaction with these meter locations is unrealistic, expensive and devoid of benefits*" Some experts advice that local governments learn from the smart meter roll out experiences of other governments in order to avoid mistakes like installing devices that, "*didn't provide consumers with real-time consumption data*"~Item #34.

3.3.3 Sustainability in the Energy Data Policy Mix

The current energy transition is commonly described as a sustainability transition. Energy policy has significant sustainability implications by shaping the pathways for energy production, consumption, and environmental protection. Effective energy policies can drive decarbonization, improve energy security, and ensure equitable and inclusive outcomes, although poorly designed policies risk perpetuating fossil fuel dependence, increasing environmental harm, and exacerbating social inequalities. However, challenges such as policy enforcement, financing, and coordination often hinder the immediate effectiveness of energy policies, especially in developing countries, emphasizing the need for integrated approaches and climate finance to achieve sustainability goals.

Many sustainability transitions scholars acknowledge (and/or point to) the need for a variety of policy instruments to achieve sustainability goals (Grin et al. 2010; Kern et al. 2019) and to avoid institutional and policy failures in the course of a transition (Weber and Rohracher 2012). Sustainability as both a technical and normative goal within energy data governance. In terms of the former, integrating data-driven insights can optimize energy efficiency, reduce emissions, and manage renewable integration. In this sense, energy data can have positive sustainability effects. At the same time, energy data is Janus-faced: it involves sustainability challenges such as the environmental costs of data infrastructure and the sustainability implications of different governance models related to energy data.

During sustainability transitions, the risk that a policy mix will be inherently incompatible is high (Edmondson et al. 2019; Flanagan et al. 2011). This is because sustainability cuts across multiple policy domains from agriculture to economy, involves high level of uncertainty and contestation, and generates externalities—both positive and negative—for constituencies in distant geographies and time-periods (Kern et al. 2019; Long et al. 2023). Indeed, the complexity and duration of transitions (Geels 2002; Geels et al. 2017) requires the development and adoption of multiple policy instruments, each serving multiple objectives (Edmondson et al. 2019; Kern and Howlett 2009; Loorbach and Rotmans 2010).

Energy data is critical in shaping effective energy policies by providing detailed, timely, and localized information that informs decision-making across multiple scales. It enables better energy efficiency planning, resource allocation, and impact assessment while supporting transparency and stakeholder engagement. Data from energy consumption patterns, infrastructure performance, and renewable integration help tailor policies to local needs, mitigate risks, and ensure equitable outcomes. Moreover, energy data facilitates monitoring progress toward climate and sustainability goals and supports the integration of transition minerals and energy-intensive sectors like data centers into coherent policy frameworks. Therefore, robust, accessible, and well-governed energy data systems are essential for enabling just and sustainable energy transitions.

In the media articles, energy data is often viewed as a tool for sustainable policy. The nested nature of energy data in sustainability policy can be seen in Singapore's policy on green buildings. This sustainability policy is heavily subsidized (supported by a $100 million SGD (approx. $790,000 USD) Green Mark incentive scheme complemented by the Singapore Building Retrofit Energy Efficiency Financing (BREEF) scheme) to incentivize building owners to bring their buildings "up to greening standards" as outlined by newly introduced legislation. Energy data played a role in this policy mix as the new law requires, "*nothing less than a three-year energy audit on building cooling systems, an annual submission of building information and energy consumption data by building owners*"~Item #23. Energy data is also part of the advocacy around greening school buildings in the U.K. where environmental organizations have adviced, "*practical changes for curriculum ideas, from developing sculpture trails, to recording energy consumption data and researching renewable resources.*"~Item #22. Indeed, weak or insufficient energy data is often seen as indicator that a project is not as green as it claims to be. For instance, in one article on green buildings, an interlocutor is quoted as saying, "*Quite a few of these green buildings turn out to be dogs from an energy consumption standpoint. Most green buildings I see getting press coverage fail to publish energy consumption data*"~Item #15. Indeed, the media articles suggest that governments use energy data to measure policy performance across a range of climate policies ~Item #9.

The prioritization of sustainability not only changes energy data production, it also changes how it is shared, by whom and to whom. The push for green energy has led to energy efficient streetlighting using smart grid technology in Florida. The private company that manages the network "*struck a deal in Florida to use its technology to transfer energy consumption data from individual homes through its network of streetlights to the utilities which, in turn, pay for the service*"~Item #21.

3.3.4 Social Marginalization in the Energy Data Policy Mix

The energy transition, although essential for climate goals, can exacerbate energy poverty and social marginalization, particularly among black, indigenous, and people of color (BIPOC) and other marginalized communities. Rising energy costs and uneven access to renewable energy subsidies often disproportionately impact low-income households, limiting their ability to participate fully in the transition and increasing social inequality. Energy data plays a crucial role by informing targeted policies that recognize and address these disparities, enabling more inclusive participation and resource allocation. In the same way that energy transitions have their own winners and losers, so too does energy data. There are biases in data access, representation, and usage that can deepen existing inequalities.

BIPOC communities can be further marginalized by energy data in several ways. First, inadequate or biased data collection may omit or misrepresent the energy needs and consumption patterns of BIPOC households, leading to policies that overlook their specific challenges. The digital divide and limited data literacy further reduce these communities' ability to control their data, increasing vulnerability to exploitation and exclusion from benefits like clean energy programs. Ensuring data sovereignty, transparency, and community-led governance is essential to protect privacy rights and foster energy justice in the energy transition.

Second, unequal access to energy efficiency programs and clean energy technologies is often perpetuated by data-driven incentive structures that favor homeowners or affluent groups, excluding many BIPOC renters and low-income households. One media article from the U.K. observes, "*Smart meters have also been found not to work in some rural areas and buildings (flats for example) where the wireless network they depend upon—operated by the Data Communications Company (DCC)—is unable to send energy consumption data to suppliers. Also, many households still have smart meters that stopped working when they switched energy supplier*"~Item #5.

Third, data privacy concerns and the digital divide can limit BIPOC communities' ability to engage with and benefit from smart energy systems and programs, reinforcing exclusion. Here, energy data can also be painted as a solution. For instance, the COVID-19 pandemic heightened awareness of energy poverty. A project that responds to this need with the aim of increasing access to energy in Detroit, Michigan. The project will, "*collect anonymized energy-consumption data from smart meters in all of the homes...to develop algorithms that identify consumers' basic electricity needs. The researchers will also weigh insights from the focus groups, where residents will be asked about their electricity-dependent household necessities*"~Item #2. The project is designed to be participatory so that the use of energy data does not perpetuate historical injustices by embedding systemic inequities within algorithms, decision-making, and policy design, ultimately reinforcing structural energy poverty and social marginalization for BIPOC populations.

And finally, their data rights may be violated. For instance, energy companies are described as using deception and manipulation to induce costumers to switch to smart meters: "*Trading standards chiefs have complained that energy firms are misleading customers by wrongly saying the devices are a legal requirement*"~Item #19. In another case, a reporter from the Daily Mail describes the energy data relationship between energy producers and consumers in this way: "*Energy suppliers have been regularly criticised for bullying households into adopting smart meters by claiming their existing meters have expired or must be replaced as a matter of urgency*"~Item #5. Those who do not comply, the article further claims, are framed as "rejectors" by energy companies as a way to attribute policy failure to these individuals~Item #5. The stereotyping of energy consumers also happens elsewhere as documented in an article from the U.S. ~Item #29. Contrary to the depiction of these individuals as backward and anti-progress that the energy companies try to push is the negative experience that many marginalized communities have had with smart meter technology. A media article describes homeowners' resistance as rooted in, "*possible negative health effects and the intrusion of privacy*" as well as claims that the meters are, "*inaccurate, overcharge consumers and pose serious health and privacy concerns.*"~Item #29. Although the concerns about the health consequences of smart meters use of radiofrequency transmissions are not validated, there is much more evidence to support the claim about privacy threats related to smart meters.

3.3.5 Privacy in the Energy Data Policy Mix

Energy data privacy poses significant risks by potentially exposing sensitive information without informed consent or adequate protections. This can lead to discriminatory outcomes such as exclusion from energy services, surveillance, and misuse of data in ways that reinforce environmental injustices and social inequalities.

The collection and use of detailed energy consumption data through smart meters and other technologies can expose consumers to heightened surveillance, misuse of personal information, and discriminatory practices, especially when data governance lacks transparency and community control. Citizens' advice groups are also raising the alarm on smart meters and privacy. For instance, one commentator describes smart meters as "*modern-day Trojan horses' that could harvest vast amounts of data about our activities*" and further voice contempt about how, "*There is already proof that data from smart meters is being monetised*"~Item #19. Limited access to data literacy and digital resources their ability to understand and manage their own data, often resulting in reduced agency and increased vulnerability. This can lead to exclusion from energy programs or unfair pricing, reinforcing social and economic marginalization.

Additionally, people living in developing countries face significant challenges related to energy data privacy and sovereignty, often compounded by weak regulatory frameworks and limited enforcement. As digital technologies and data-driven energy systems expand, these populations risk exposure to data misuse, surveillance, and lack of control over personal information, which can exacerbate vulnerabilities and exclusion from energy services.

Addressing these issues requires strengthening legal frameworks, increasing public education on data privacy, and promoting inclusive data governance that reflects the needs and rights of vulnerable communities in the energy transition. Consequently, privacy has made energy data a target of policy in some places. For example, in British Colombia, Information and Privacy Commissioner provided oversight over the regions' smart meter programs after, "*after receiving numerous complaints that the information collected by the meters may breach personal privacy*"~Item #39. However, as we will see in Chap. 6, the efforts to protect consumer privacy are fragmented and insufficient.

3.3.6 Problem Shifting Through an Ecology of Games

Energy data is regulated through a policy mix of energy policy instruments and digitalization policy instruments. Without careful design and inclusive governance, the transition risks deepening inequalities and exclusion, underscoring the importance of embedding justice, recognition, and equitable access at the core of energy policy to avoid worsening energy poverty and marginalization. Current energy data policy is lacking. Its overreliance on voluntary initiatives by energy providers can produce substantial implementation gaps. When the policies, laws, or regulations across different organizations, governments, or jurisdictions are misaligned, the result is miscoordination, inconsistencies, conflicts, overlaps, inefficiencies, inefficacies, and injustices in the current system.

Interactions between policies in the mix are expected to be harmonious when policy mixes are complementary (Kern and Howlett 2009); however, policy mixes may also be inherently incompatible (Flanagan et al. 2011). This is because at any given moment, policy making involves a network of actors working in different policy venues (Jenkins-Smith et al. 2018; Kingdon 2003), some of these actors work in multiple venues which differ based on issue area (e.g., in battery policy and in labor policy), geographic scale (e.g. local or cross-border governance), and/or legal mandate (Sabatier). The interests and aims of actors do not always align and competition amongst them can emerge (Berardo and Lubell 2019; Jenkins-Smith et al. 2018). This encourages actors to strategically move their advocacy from one policy venue to another (e.g., moving from the national platforms to the E.U. to find a more "welcoming" policy ecosystem (Guiraudon 2000)), they may create positive or negative externalities in the other venues that they engage in (Berardo and Lubell 2019; Lubell 2013).

The externalities in the policy mix may be payoff externalities. Payoff externalities are the collective action implications for policy issue A (e.g., in energy policy) that result from decision-making on policy issue B (e.g., in digitalization policy). This can happen, for instance, when two different rule sets (or institutions) are applied in the same jurisdiction (e.g., policies on data production for energy policy and on data production for digitalization) (Lubell 2013). This can be confusing and the resulting blind spots have been associated with spillover effects and unintended consequences. One of those adverse side effects may be problem shifting (Wood Hansen and Van Den Bergh 2023).

Energy data straddles two domains: energy policy and digitalization policy; therefore, it is important to consider policy mobility in both fields. Because energy data is shaped by policy making in both areas, we must consider how misalignment between these areas can be shaped by the biophysical, economic, or social processes that link these two issue spaces (Lubell 2013). What results is an ecology of games. The theorizing on the ecology of games (Long 1958) takes as a point of departure the assumption that many policy problems are coupled (e.g. land use and energy) (Berardo and Lubell 2019; Wood Hansen and Van Den Bergh 2023). It follows that the nature of these couplings shapes the efficacy and ethics of policymaking. This can be seen in the case of energy data.

The coupling of two policy areas like energy and digitalization can lead policymakers to have an incomplete understanding of the energy data and may also lead to policy fragmentation. Here, the concept of policy mixes can be helpful. Policies and their objectives are often iteratively nested and/or connected over time to become embedded in policy mixes (Flanagan et al. 2011; Kern et al. 2019). This is especially likely in the case of innovation policy because they are often mainstreamed into the agendas of many traditional policy fields. As such the focus of energy data policy analyses need to look beyond specific policies and at the interaction between different policies that are affected by energy data (Flanagan et al. 2011; Nauwelaers and Wintjes 2002). Without the application of a holistic approach, the likelihood of problem shifting and failure of energy data policy is substantially higher.

3.4 Conclusion: Toward Integrated Energy Data Governance

The objectives and visions pursued through policy instruments are entangled with a number of social values (Flanagan et al. 2011). For instance, energy data is necessary for enabling energy efficiencies (e.g., by improving energy reliability) (Koseleva and Ropaite 2017; Lewis et al. 2020), assist in the development of energy technologies (Agrawal and Loon 2023), and promote energy justice (e.g. through the detection and reduction of energy poverty in hard to serve areas (Branscomb and Florida 1998; Ramírez 2021).[2] However, as depicted in Fig. 7.1, energy data is Janus-faced: sustainability, social marginalization (justice), and privacy are deeply interconnected concerns within energy data governance. Not only can the production, sharing, and storage of energy data run counter to the achievement of these social

[2] A more extensive discussion of energy data promises and potentialities can be found in Chap. 1.

goals[3]: these social goals can share both complementary and contradictory relationships amongst themselves and thus call for new governance alternatives. Building on this understanding, this chapter examines how misalignments within the mix of policy governing energy data or affected by energy data governance can give rise to problem shifting, which in turn produce tensions between the social goals that are connected to energy data.

An understudied and poorly understood dimension of policy mixes is its justice implications. This is what the problem shifting lens brings to the fore. When a policy mix generates losses for a stakeholders or domains, the expectation is that they will actively oppose the mix (Edmondson et al. 2019). However, when it comes to energy data the lion's share of the benefits befall private actors like energy companies and third party intermediaries, while the burdens—especially related to securing energy data—are diffuse making it these negative consequences less perceptible, and increasing the difficultly for an opposition force to mobilize and coordinate (Lowi 1972; Olson 2003). Moreover, even when the identity of those on the losing end is very specific and visible (e.g., those whose access to drinking water is affected by data centers), that constituency may lack the power or the political support to oppose a policy mix (schneider and ingram). As a result, the development and use of innovations, like energy data can create injustices that are poorly addressed by socio-technical regimes.

Appendix

See Table 3.2.

References

Agrawal TK, Loon PV (2023) Vehicle and stationary batteries for a sustainable future: recent developments and challenges around second-life applications and recycling1. In Urban F, Nordensvärd J (eds) Handbook on climate change and technology. Edward Elgar Publishing, pp 291–301. https://doi.org/10.4337/9781800882119.00031

Ardoin PJ, Grady D (2006) The politics of electricity restructuring across the American states: power failure and policy failure. State Local Govern Rev 38(3):165–175. https://doi.org/10.1177/0160323X0603800304

Baldwin E, Carley S, Nicholson-Crotty S (2019) Why do countries emulate each others' policies? A global study of renewable energy policy diffusion. World Dev 120:29–45. https://doi.org/10.1016/j.worlddev.2019.03.012

Berardo R, Lubell M (2019) The ecology of games as a theory of polycentricity: recent advances and future challenges. Policy Stud J 47(1):6–26. https://doi.org/10.1111/psj.12313

Borrás S, Edquist C (2013) The choice of innovation policy instruments. Technol Forecast Soc Chang 80(8):1513–1522. https://doi.org/10.1016/j.techfore.2013.03.002

[3] See Chaps. 4, 5, and 6 of this book to learn more about each of these aspects.

Branscomb LM, Florida R (1998) Challengesto technology policy in a changing World
Cairney P (2012) Understanding public policy: theories and issues. Palgrave Macmillan.
Capano G, Howlett M (2020) The knowns and unknowns of policy instrument analysis: policy tools and the current research agenda on policy mixes. SAGE Open 10(1):2158244019900568. https://doi.org/10.1177/2158244019900568
Cia Alves EE, Steiner A, De Almeida Medeiros M, Da Silva MEA (2019) From a breeze to the four winds: a panel analysis of the international diffusion of renewable energy incentive policies (2005–2015). Energy Policy 125:317–329. https://doi.org/10.1016/j.enpol.2018.10.064
Dekker R, Van Est R (2020) The convergence of electricity and digitalization in the Netherlands: data governance as an emerging policy issue. TATuP Zeitschrift Für Technikfolgenabschätzung in Theorie Und Praxis 29(2):31–37. https://doi.org/10.14512/tatup.29.2.31
Edmondson DL, Kern F, Rogge KS (2019) The co-evolution of policy mixes and socio-technical systems: towards a conceptual framework of policy mix feedback in sustainability transitions. Res Policy 48(10):103555. https://doi.org/10.1016/j.respol.2018.03.010
Elo S, Kyngäs H (2008) The qualitative content analysis process. J Adv Nurs 62(1):107–115. https://doi.org/10.1111/j.1365-2648.2007.04569.x
Flanagan K, Uyarra E, Laranja M (2011) Reconceptualising the 'policy mix' for innovation. Res Policy 40(5):702–713. https://doi.org/10.1016/j.respol.2011.02.005
Geels FW (2002) Technological transitions as evolutionary reconfiguration processes: a multi-level perspective and a case-study. Res Policy 31(8–9):1257–1274. https://doi.org/10.1016/S0048-7333(02)00062-8
Geels FW, Sovacool BK, Schwanen T, Sorrell S (2017) The socio-technical dynamics of low-carbon transitions. Joule 1(3):463–479. https://doi.org/10.1016/j.joule.2017.09.018
Gill GS (1976) Perverse economic incentives and energy conservation. Energy 1(4):445–450. https://doi.org/10.1016/0360-5442(76)90072-4
Grin J, Rotmans J, Schot J (2010) Transitions to sustainable development: new directions in the study of long term transformative change. Routledge. https://doi.org/10.4324/9780203856598
Guiraudon V (2000) European integration and migration policy: vertical policy-making as venue shopping. JCMS J Common Mark Stud 38(2):251–271. https://doi.org/10.1111/1468-5965.00219
Hoppmann JH (2013) The role of deployment policies in fostering innovation for clean energy technologies: Insights from the solar photovoltaic industry. ETH Zurich, Application/pdf, p 1 Band. https://doi.org/10.3929/ETHZ-A-009787261
Howlett M, Mukherjee I, Woo JJ (2015) From tools to toolkits in policy design studies: the new design orientation towards policy formulation research. Policy Polit 43(2):291–311. https://doi.org/10.1332/147084414X13992869118596
Jenkins-Smith HC, Nohrstedt D, Weible CM, Ingold K (2018) The advocacy coalition framework: an overview of the research program. In: Weible CM, Sabatier PA (eds) Theories of the policy process, 4th edn. Routledge, pp 135–171. https://doi.org/10.4324/9780429494284-5
Karch A, Nicholson-Crotty SC, Woods ND, Bowman AO (2016) Policy diffusion and the pro-innovation bias. Polit Res Q 69(1):83–95. https://doi.org/10.1177/1065912915622289
Kashyap S, Schwinger W, Grünberger S, Schaffer C, Hartner G (2025) Energy Data governance fragmentation in europe: challenges, models, and pathways to integration. In: 2025 21st International Conference on the European Energy Market (EEM) (pp 1–6). https://doi.org/10.1109/EEM64765.2025.11050331
Kern F, Howlett M (2009) Implementing transition management as policy reforms: a case study of the Dutch energy sector. Policy Sci 42(4):391–408. https://doi.org/10.1007/s11077-009-9099-x
Kern F, Rogge KS, Howlett M (2019) Policy mixes for sustainability transitions: new approaches and insights through bridging innovation and policy studies. Res Policy 48(10):103832. https://doi.org/10.1016/j.respol.2019.103832
Kim JE (2020) Regulation trumps economics? Examining renewable energy policy, diffusion and investment in 80 developing countries. Energy Res Soc Sci 70:101613. https://doi.org/10.1016/j.erss.2020.101613

Kim RE, Van Asselt H (2016) Global governance: problem shifting in the Anthropocene and the limits of international law. In: Morgera E, Kulovesi E (eds) Research handbook on international law and natural resources. Edward Elgar Publishing. https://doi.org/10.4337/9781783478330.00039

Kingdon JW (2003) Agendas, alternatives, and public policies, 2nd edn. Longman

Koseleva N, Ropaite G (2017) Big data in building energy efficiency: understanding of big data and main challenges. Procedia Eng 172:544–549. https://doi.org/10.1016/j.proeng.2017.02.064

Kyngäs H (2020) Inductive content analysis. In: Kyngäs H, Mikkonen K, Kääriäinen M (eds) The application of content analysis in nursing science research (). Springer International Publishing, pp 13–21. https://doi.org/10.1007/978-3-030-30199-6_2

Latour B (ed) (2005) Reassembling the social: an introduction to actor-network-theory. Oxford University Press

Lewis J, Hernández D, Geronimus AT (2020) Energy efficiency as energy justice: addressing racial inequities through investments in people and places. Energ Effi 13(3):419–432. https://doi.org/10.1007/s12053-019-09820-z

Long NE (1958) The local community as an ecology of games. Am J Sociol 64(3):251–261. https://doi.org/10.1086/222468

Long LAN, Krause RM, Arnold G, Swanson R, Fatemi SM (2023) The networked micro-decision context: a new lens on transformative urban governance. Urban Transform 5(1):9. https://doi.org/10.1186/s42854-023-00054-y

Loorbach D, Rotmans J (2010) The practice of transition management: examples and lessons from four distinct cases. Futures 42(3):237–246. https://doi.org/10.1016/j.futures.2009.11.009

Lovell H, Nixon C, Betzold A (2023) Policy mobilities and the policy cycle: an analysis using two smart grid case studies. Geoforum 144:103818. https://doi.org/10.1016/j.geoforum.2023.103818

Lowi TJ (1972) Four systems of policy, politics, and choice. Public Adm Rev 32(4):298. https://doi.org/10.2307/974990

Lubell M (2013) Governing institutional complexity: the ecology of games framework. Policy Stud J 41(3):537–559. https://doi.org/10.1111/psj.12028

Makse T, Volden C (2011) The role of policy attributes in the diffusion of innovations. J Polit 73(1):108–124. https://doi.org/10.1017/S0022381610000903

McDonnell LM, Elmore RF (1987) Getting the job done: alternative policy instruments. Educ Eval Policy Anal 9(2):133–152. https://doi.org/10.3102/01623737009002133

Nauwelaers C, Wintjes R (2002) Innovating SMEs and regions: the need for policy intelligence and interactive policies. Technol Anal Strat Manag 14(2):201–215. https://doi.org/10.1080/09537320220133866

Nguyen Long LA, Jansma SR, Lee D (2024) Changing the game: situating energy citizenship in the Dutch socio-technical landscape. J Clean Prod 483:144258. https://doi.org/10.1016/j.jclepro.2024.144258

Nicholson-Crotty S, Carley S (2016) Effectiveness, implementation, and policy diffusion: or "can we make that work for us?" State Polit Policy Quart 16(1):78–97. https://doi.org/10.1177/1532440015588764

Olson M (2003) The logic of collective action: public goods and the theory of groups (21. printing). Harvard University Press

Patashnik E, Zelizer JE (2009) When policy does not remake politics: the limits of policy feedback. In: APSA 2009 Toronto Meeting Paper. SSRN https://ssrn.com/abstract=1449996

Peck J, Theodore N (2015) Fast policy: experimental statecraft at the thresholds of neoliberalism. University of Minnesota Press

Pierson P (1996) The new politics of the welfare state. World Polit 48(2):143–179

Ramírez C (2021) Without environmental justice, the renewable energy transition will leave low-income and BIPOC communities behind. J Sci Policy Govern 18(03). https://doi.org/10.38126/JSPG180306

Richards D, Smith MJ (2008) Governance and public policy in the United Kingdom (Reprinted). Oxford University Press

Rosenow J, Eyre N (2016) A post mortem of the green deal: austerity, energy efficiency, and failure in British energy policy. Energy Res Soc Sci 21:141–144. https://doi.org/10.1016/j.erss.2016.07.005

Schaefer M, Schmitt Olabisi L, Arola K, Poitra CM, Matz E, Seigel M, Schelly C, Adesanya A, Bessette D (2021) Understanding socio-technological systems change through an indigenous community-based participatory framework. Sustainability 13(4):2257. https://doi.org/10.3390/su13042257

Schneider A, Ingram H (1990) Behavioral assumptions of policy tools. J Polit 52(2):510–529. https://doi.org/10.2307/2131904

Shipan CR, Volden C (2008) The mechanisms of policy diffusion. Am J Polit Sci 52(4):840–857. https://doi.org/10.1111/j.1540-5907.2008.00346.x

Smith K (1994) New directions in research and technology policy: identifying the key issues. STEP Report series 199401, The STEP Group, Studies in technology, innovation and economic policy

Sokołowski MM, Heffron RJ (2022) Defining and conceptualising energy policy failure: the when, where, why, and how. Energy Policy 161:112745. https://doi.org/10.1016/j.enpol.2021.112745

Wood Hansen, O., & Van Den Bergh, J. (2023). Environmental problem shifting from climate change mitigation: a mapping review. PNAS Nexus 3(1):pgad448. https://doi.org/10.1093/pnasnexus/pgad448

Yoon J-H, Sim K (2015) Why is South Korea's renewable energy policy failing? A qualitative evaluation. Energy Policy 86:369–379. https://doi.org/10.1016/j.enpol.2015.07.020

Part II
Who Gets What, When, and How in Energy Data?

Chapter 4
Energy Data Production: Opportunities and Challenges

4.1 Households as Data Generators

Although energy systems generate an array of data, the focus of this book is on energy consumption data. Energy end users are any entity that consumes energy. Energy is a necessary resource for a range of human activities from leisure to food production (Mousavi-Avval et al. 2011), all of which manifests as energy consumption and relevant energy data. Public schools, for instance, consume energy in their day-to-day operations. Political and social developments can place pressure on this relationship. For instance, climate change creates a political and material urgency around improving the energy performance of public school buildings, and induce efforts to track and benchmark the energy consumption of school buildings (Dias Pereira et al. 2014). In this chapter, our specific focus will be on energy data produced by households. We can understand these energy data's life cycle as it spans from production to knowledge generation and includes acquisition, processing, integration, and mining as intermediary steps (Wang et al. 2023). These data are generated via household's energy consumption and consequent financial exchanges, for instance payment records (Lee et al. 2019). As such, the choice to focus on consumption emphasizes the role of energy end users as producers of energy data.

Energy data is both valuable and not value-neutral, and they stimulate and support a status quo which favors the energy producers. Although their use is constrained by legal frameworks such as the E.U.'s General Data Protection Regulation (GDPR) that have codified rights that empower consumers to access their data, consent to its use, and restrict unwanted sharing, coverage and implementation of such regulations is often incomplete and fragmented, leading to the frequent abrogation and abuse of energy consumers data rights. Best practices advocate for transparent communication from utilities and service providers with the goal of ensuring that consumers understand what data is collected, who can access it, and for what purposes. Experts argue that consent mechanisms should be user-friendly, clearly opt-in, and allow individuals to amend permissions as desired. Despite these existing and proposed protections,

L. A. Nguyen Long and D. Lee, *Energy Data*, Lecture Notes in Energy 46,
https://doi.org/10.1007/978-3-032-19171-7_4

considerable challenges remain in practical implementation, raising the likelihood that energy data will be a font for political contestation. A first step in addressing this implementation gap is to examine the data production process. Before that, we need to establish a common understanding for the term, energy data.

4.1.1 Energy Data Contain Multitudes

Whether the end user is a household or company, energy data can help develop evidence-based and effective solutions to inefficient and unsustainable energy use without adversely impacting quality of life or productivity for end users, and revenue streams for utility companies. Energy is implicated in three core challenges facing our generation: climate change, (energy) poverty, and (energy) insecurity. Diversifying the energy mix and decentralizing energy through renewable energy is one part of the solution. Enhancing energy performance by raising energy efficiency may be another. To achieve both ends, energy data is needed to design interventions and technologies which can enable decentralization (Williams and Short 2020; Wu et al. 2021), raise consumer (households and industries alike) awareness and ameliorate energy consumption behaviors (Abele et al. 2015; Shrouf and Miragliotta 2015).

Household consumption increasingly accounts for a larger share of overall energy consumption (Zhou and Yang 2016). In the E.U., household energy consumption accounted for 26.2% of final energy consumption across the territory in 2023 (Eurostat 2025). In the U.S. residential energy use accounted for 19.7% of total energy consumption in that same period (US Energy Information Administration 2024). At the same time, the share of total energy consumption attributed to household is growing in emerging economies like China (Tao et al. 2018; Zheng et al. 2024; Zou and Luo 2019) and India (Thapar 2020).

Households produce massive amounts of energy data through their diverse uses (and non-uses) of energy, and different energy mixes are used to meet their energy needs. For instance, there is a marked difference in the types of energy (e.g., natural gas versus renewable energy) used in the U.S. and the E.U., or between nation states in the E.U. (Eurostat 2025). Even inside of a country, demographics not only determines the type of energy used but also data availability. In the U.S., energy-poor communities often lack the information and communications technology needed to support the production and transmission of fine-grained energy system performance data (Lee et al. 2019). This suggests that there is significant within and between country variation in how and what energy is being consumed by households.

Heterogeneity in energy data is driven by diverse factors. Zhou and Yang (2016) specify three dimensions of household energy consumption behavior that can help unpack this variation. The first dimension is time—the period (e.g., hours, days, weeks, months, years) within which consumption is measured, and patterns are observed. The authors observe that the longer the time span of data collection is, the less the variation is likely to be due to random chance (Zhou and Yang 2016). The second dimension is the spatial dimension—external environment and climate, building structure, etc.—affecting energy needs. Given the volume and nuance of this data, it can often be considered "big data."

Zhang et al. (2018) describe big data as "a collection of data sets that are too large and complex to efficiently manage and process using the traditional technologies and tools" (Zhang et al. 2018: p. 4). These data are characterized by high volume—i.e., large amounts of data, high velocity—i.e., the rapidity with which speed of energy data is collected, processed and analyzed, high variety—referring to the different mixes of structured, semi-structured, and unstructured data collected, and high value for consumers and utility companies, which belong to a typical family of big data (Koseleva and Ropaite 2017; Zhou and Yang 2016). In addition to these parameters of big data outlined above (e.g., the different types of data collected and their volume, velocity, etc., the technical design of the data generating system (spanning collection, processing, and analysis), the level of "completeness" of the data, and the legal mores that are applicable to said data (e.g., personal identifiable information versus aggregate data), are all features that have—within the scholarly literature—been associated with the notion of data (Hummel et al. 2021).

We can expect the energy needs of households in temperate climates to be markedly different from households in more extreme climates. In the United States, the least electricity is consumed each year by apartments in the Northeast consume while single-family detached homes in the South demonstrate the most substantial energy use (US Energy Information Administration 2023). Finally, there is the energy user themselves—or the internal (subjective, e.g. habits) and external (e.g., occupation, demographic characteristics, etc.) affecting consumption behaviors. For example, the intensity of energy use may depend on the types of occupations and lifestyles of an individual. This user-centered difference may explain the marked difference in the types of energy resources used in urban and rural households, for instance in China (Tao et al. 2018; Zou and Luo 2019). This temporality, spatiality, and domain-centric view of energy consumption echo the dimensionality of the problem shifting perspective. Later in the chapter we explore how these inputs are amplified by problem shifting when situated within a process of irresponsible innovation.

4.1.2 How Smart Meters Are Changing the Energy Data Production Game

Data is a technology which requires other technologies—a fact which complicates the achievement of responsible innovation. To begin, the data production architecture must be sufficiently sophisticated to capture the energy data in all its nuance and richness. After all, as explained above, energy data involves heterogeneous end-users—as reflected in the diverse combination of dimensions embodied by different users—, energy mixes, and big data types. The collection of these energy data brings a complex of technology into the fold. This is the crux within which responsible innovation and problem shifting dynamics must be interrogated: the architecture enabling data collection through sensing and measurement technology, Information and Communications Technologies (ICTs), and big data storage and analytics (Zhou and Yang 2016) is rife with opportunity and the actuality of irresponsible organization. Given the legion of actors, technologies, and data involved in this data generation architecture, fragmentation is the rule rather than the exception. In such a data production architecture, smart meters are critical devices for data production.

The adoption of smart meters represents a transformative leap in energy data production and collection. Traditional analog meters required monthly manual readings, which could be time consuming and error prone. In contrast, smart meters provide real-time, granular data along several parameters (e.g., power consumption, factor, and peak voltage, phase angle and frequency, the meter identifier, the data timestamp) on electricity use at the household or business level (Jixuan Zheng et al. 2013; Depuru et al. 2011; Shrouf and Miragliotta 2015). The energy data on household consumption are collected from those end users and transmitted to utility companies (Armoogum and Bassoo 2019; Jixuan Zheng et al. 2013).

The shift from analog meters to smart meters has enabled utility companies to monitor usage patterns by the hour or even minute, optimizing grid management, fault detection, and demand forecasting (Forbush 2012; Depuru et al. 2011). When consumers access their data, they can in theory use it to gain more immediate insights into their consumption and adopt energy-saving behaviors. In comparison to households, companies more frequently make use of the energy data generated by smart meters. Companies have greater capacity to deploy energy data strategically and purposefully. The three levels of smart meters in industry span the factory-level, production line-level, and machine level, where most of the fine-grained data are produced (Kara et al. 2011).

Smart meters are not simply technological devices, they create knowledge systems (Zhang et al. 2018), and thus implicate questions of power and agency. Energy data, like all types of knowledge contribute to the design, maintenance, and governance of sociotechnical systems, like an energy system. Knowledge, how it is produced, distributed, used, and even discarded reflects the various contexts, interests, and institutions that drive system dynamics. It is a source of power for various system stakeholders, and helps define the roles, inclusion, and possibilities for system actors and technologies (Foucault 2020; Weckert 2017). It is in this way that knowledge and social order are co-produced, and co-constitutive (Jasanoff 2010).

The systems that smart meters initiate span hardware including the smart meter itself, communication infrastructure, sensors and control devices alongside software techniques and data processing (Depuru et al. 2011). Moreover, smart meters are a key component of smart grids. Unlike traditional power grids, which operate centralized power distribution systems, smart grids enable power flow and two-way information flow for decentralized distribution (Jixuan Zheng et al. 2013). Smart meters leverage the resulting architecture to support decentralized energy generation through bidirectional exchange of electricity and information between utility companies and their customers, they are also able to communicate with other meters that are within their reach (Jixuan Zheng et al. 2013; Zhang et al. 2018).

One thing implied by bidirectional communications involving smart meters is the need for an effective communications network. Key features of an ideal smart meter network are cost-effectiveness, reliability and security (Depuru et al. 2011). Reliable networks provide constant support to the smart meter system operations, even during power outages, while the ability to prioritize the data delivery to different units increases network cost effectiveness. Given the substantial amount of (at times sensitive and confidential) data being transferred and the diverse units that the smart meter communicates with to obtain and share data (e.g., utility company and home appliances) which means that both human, physical, and digital components of the network must operate in adherence to security guidelines regarding the transmission, collection, storage, and maintenance of energy data (Depuru et al. 2011). Working from within this network smart meters provide real time energy data, monitor energy storage devices, bill customers, execute control commands locally and remotely, for instance by monitoring and controlling household appliances (Depuru et al. 2011).

Producing energy data with smart meters is not without its challenges. The complexity, volume, and speed of big energy data calls for increased computational capacity (Zhang et al. 2018), which in turn calls for scalable information systems. Ceci et al. (2014: p. 64) describe scalability as "the ability of a system to handle a growing amount of information in an efficient and effective way." It involves the ability to rapidly adapt and upgrade hardware and computing facilities. They further argue that a system needs to be able to absorb malfunctions and mistakes, a feature they refer to as *fault-tolerance* (Ceci et al. 2014). Maintaining and keeping these systems up to date may be costly for energy utilities companies (Depuru et al. 2011).

Energy data collected using smart meters also demand larger and more data centers for data storage. And, data centers involve their own energy efficiency challenges. For instance, Dayarathna et al. (2015) estimate that "a typical data center may consume as much energy as 25,000 households. Data center spaces may consume up to 100 to 200 times as much electricity as standard office space. Furthermore, the energy costs of powering a typical data center doubles every five years." This energy use has been linked to numerous environmental issues, despite the industries' efforts to increase energy efficiency and to reduce reliance on fossil fuels (Koseleva and Ropaite 2017).

Agency, control, power and autonomy are not evenly distributed across smart meter information systems. This imbalance creates opportunities for misuse and abuse of energy data. Clearly, there is a need for responsible innovation principles to guide the further development and implementation of energy data technologies like smart meters. A first step in this direction is to acknowledge the complexity of energy data itself.

4.2 Energy Data: More Than Just Data

Energy data is not neutral. Moreover, different actors associate different advantages to energy data. The wealth of data provided by smart meters carry significant benefits and risks for different stakeholders, all of which are distributed unevenly. For instance, energy data holds distinct potential benefits (e.g. energy efficiency) and risks for energy producers—or utility companies—for governments and policymakers, and for end users like households and companies.

A framework that can be instructive for understanding energy data production is Bruno Latour's Actor-Network Theory (ANT), which argues that humans and nonhumans (technologies, artifacts) form networks (Latour 2005; Rowland et al. 2011; Valverde 2007). These networks are constituted of relationships between actors and technology involving active mediation and mutual influence. Technologies—like smart meters, grids, and solar panels—are understood to be non-human actors that participate actively within networks, shaping and being shaped by human and social interactions (Latour 2005; Rowland et al. 2011). Within this assemblage, technologies can shape behaviors and meaning making. In this sense, innovation can be understood as a process of alignment involving negotiation, power, and translation by heterogeneous actors who interact iteratively. Within this dynamic interaction, foresight and reflexivity are key. It is important to anticipate the consequences of deepening human to technology interactions to prevent unhealthy dependencies or misuses of technologies.

When reflecting on energy data consequences—both current and future—it is important to recognize that the human actors in energy systems are different from one another and that they all derive diverse benefits from energy data and confront different burdens and risks. Some actors even have the power to shift burdens and risks to other actors within the system, or to actors in contiguous domains, like land use or water safety. Recognizing the different roles, resources, interests, and inequities baked into the social relations of an actor-technology network requires reflexivity—especially the consideration of ethical and political implications—and inclusivity to be able to integrate diverse perspectives. To understand the need for inclusivity and reflexiveness, let us consider the variation in the relationships that diverse stakeholders share with energy data.

Utility companies enjoy diverse benefits energy data generated by smart meters. They can leverage big energy data to enhance grid reliability, integrate renewable sources (e.g., solar and wind power) more seamlessly into the grid, manage energy production and storage more effectively, provide more efficient pricing, offer customers greater control over their energy consumption, and accelerate the transition to a greener grid (Forbush 2012; Depuru et al. 2011). Availability of fine grained data in real time increases stability and reliability by helping utility companies correctly anticipate periods of high demand, balance energy supply and demand, and avoid grid disruption and blackouts (Armoogum and Bassoo 2019). They can also generate more income by (1) using the data to produce more targeted and flexible marketing strategies (Moghaddam et al. 2011) and (2) using improved knowledge of consumption patterns to take actions that reduce enterprise risk and improve shareholder confidence (Ceci et al. 2014). In short, they can derive significant economic value from these data.

Energy data can also benefit consumers. Energy data can help companies adopt more energy-efficient practices and realize energy savings (Abele et al. 2015; Shrouf and Miragliotta 2015), but these gains are negligible compared to what other energy system actors stand to gain from energy data. Companies can use energy data (e.g., real-time energy consumption data) to reveal energy intensive devices or processes, also termed "energy sinks." This helps to identify energy-efficient technologies within a factory, and areas which can be targeted for improvement (Abele et al. 2015). In this way, companies can pursue economic and sustainability value creation by deploying energy data purposefully. Households can use personalized usage reports to adjust and optimize their electricity consumption behaviors (Zhou and Yang 2016)—and potentially lower bills. Smart meters also offer convenience and reduce the stress of behavior change by adjusting appliance use in ways that minimize energy waste (Depuru et al. 2011). Like companies, there is a potential for households to use data to derive economic and sustainability gains. These advantages may, however, be in tension with how much they value their own privacy and autonomy.

Additionally, policy makers are increasingly concerned with how to promote energy efficiency in both the public and private sectors (Zhou and Yang 2016). They can use aggregate energy data to inform policy making in ways that improve climate change mitigation (Zhou and Yang 2016). They may—for example—use energy data to facilitate decentralized energy production and allow consumers to also become small-scale producers of energy. For these policy makers, energy data can be used to increase energy security, sustainability, and also address energy poverty.

In the system described above, the status quo is maintained and power in energy systems rests largely in the hands of energy producers. Largely because they have the greatest level of control over energy data production. Benefits to end-users are minute compared to the manifold ways that utilities can leverage them, not to mention the commercialization of energy data itself. This puts into question the question of *data ownership,* which essentially is a question of inclusivity and reflexivity. Here, the contributions of Shoshana Zuboff are highly relevant (Zuboff 2020, 2015). Zuboff introduces the concept of "Surveillance capitalism" and how it strengthens asymmetries of knowledge and power between corporations and users, in our case between utilities and end-users. She observes how tech companies claim personal data as "free raw material," transforming human experiences into behavioral data for prediction and profit without meaningful oversight. This new economic logic around data, particularly consumption data, Zuboff warns readers, erodes autonomy, democracy, and privacy by concentrating knowledge and power in platforms that shape and control behavior. However, the question of control and ownership is not simply important for individual end-users, they also are important at national and international scales. The emergent discourse on data sovereignty reflects the potentially conflictual dimension of data in international relations.

Actor-network theory allows us to see that energy data is not simply static information; it functions as an actor within networks, participating in the translation and movement of information across technological and social sites. As such, energy data and the technological architecture producing it have influence over organizational practices and decision-making processes. More specifically, smart meters and smart grids are active non-human actors within the energy network that mediate how energy use is monitored, controlled, managed, and ideated. For example, smart meters can influence consumer behavior by providing real-time feedback, nudging users toward energy-saving practices. In short, these technologies shape our expectations and behaviors. They may determine our practices and what we imagine is possible in energy systems. At the same time, the smart grid integrates diverse human actors (consumers, utility companies, regulators) and technological actors (smart meters, sensors, data platforms, algorithms). All these actors co-constitutively shape the network dynamics, resulting in new forms of energy management and efficiency.

4.3 Problem Shifting in the Interest of Energy Data

To see energy data as an actor is to understand it as something which seeks to survive through reproduction and replication. The power in such a system lies in the hand of those who design processes of energy data collection and definition, which in the case of energy data is energy producers. RI reminds us that the risk inherent to systems characterized by such imbalances of power is that energy date becomes the means to its own end (which essentially serve the interests of energy producers), and that energy data can become data that is produced for data's own sake without reflection on the need for or appropriateness of further data production (Burget et al. 2017). This minimizes the agency of energy consumers, if not eliminating it entirely. In such a system of understanding and imagining energy data, the relationship between energy end-users and the system becomes extractive and energy end-users become objects from which to extract data which can then be exploited by other data users. At the heart of this power imbalance is the question: who owns energy data (we revisit the ownership of data in Chap. 7)?

Ownership and control of energy data remain contested. Is it the property of the consumer or of the technology that gathers information on that consumer and transforms it into data? In many jurisdictions, utilities claim at least partial ownership, while regulators increasingly recognize the individual consumer's rights to access, consent, and portability. With the power of what to measure, how, and how frequently, energy producers can use energy data to solve problems like energy intermittency while shifting the problem to energy consumers who now are confronted by privacy challenges. Considerations of data ownership intersect with emerging discourses on data justice (Bigo et al. 2019; Taylor 2017) and data sovereignty (Lohmöller et al. 2024), or "the more or less abstract idea that individuals, specific groups, or communities should retain control over the handling of their data" (Hummel et al. 2021: p. 5).

Data sovereignty, meaning, in this case, individuals' control over their personal data, is crucial in smart grids due to continuous collection of detailed real-time energy usage data. This data shapes grid management but raises privacy conflicts between social benefits and personal privacy. To balance these, Europe and North America apply policies like customer opt-out, incentivized opt-in, independent encrypted data storage, and dedicated privacy enforcement agencies. Despite general data privacy laws, energy sector policies face challenges balancing collective interests with individual privacy, offering models for developing more comprehensive protections tailored to energy systems (Wang et al. 2023).

The energy transitions literature is chock full of examples of environmental racism—the tendency to site problematic environmental infrastructures (e.g., incinerators or dams) within or near territories and lands where the inhabitants are predominantly black, indigenous, or people of color (Burton et al. 2024; Cranmer et al. 2023; Ramírez 2021)—and there are hints that similar patterns of wellbeing extraction is occurring when it comes to energy data (Kurz et al. 2022; Sovacool et al. 2022, 2017). If we widen the lens, we can see that energy data is commonly considered to be a matter for consideration in innovations and economic development policy. However, its production and use can lead to the creation of problems in social domains. As such, it constitutes a classic case of problem shifting. More specifically, the generation and use of energy data to increase profitability and lower emissions through greater energy efficiency can also cause social problems related to sustainability, privacy, and inequity.

4.3.1 Responsible Innovation: Social Marginalization and Data Protection Disparities

It is becoming increasingly clear that not all consumers enjoy equal protection of their energy data. Energy-poor households, renters, and those in subsidized housing may have little control over meter installation or how their data is shared. In areas lacking extensive electricity infrastructure, gaining electricity access is closely linked to increased access to information technology (Kurz et al. 2022). This expansion often marks the initial collection of personal data through sensors in newly electrified communities. Although essential for development, data collection raises ethical issues and risks that are frequently overlooked amid efforts focused on sustainable electricity expansion by governments, researchers, and utilities (Houngbonon et al. 2021; Lee et al. 2019). In short, the roll out of smart meters in some jurisdiction violates principles of RI, namely inclusion and responsiveness. In excluding vulnerable individuals from the choice of whether (or not) to install smart meters, negates their agency.

The level of agency that different households must be able to refuse the installation of smart meters may be called into question. Language barriers, digital illiteracy, and a lack of negotiating power further marginalize vulnerable groups, leaving them exposed to unconsented data use or exclusion from privacy-enhancing tools. Conversely, those with higher socioeconomic status may more readily access opt-out mechanisms or privacy settings, deepening inequalities. This creates a system where the level of responsiveness of the technology to the needs of the user depends, at least in part, to the socio-economic status of that user.

The risk that third parties will commandeer energy data goes beyond the energy producer—energy consumer relationship. Here, the question of inclusion in the overarching energy system (and therefore also in energy data systems) is once again at play. This is illustrated in the case of renewable energy technologies, which enable the consumer to also become a producer (Campos and Marín-González 2020; Jansma et al. 2023; Zafar et al. 2018) economically marginalized communities may still lose control over their energy data even when they control both energy production and energy consumption. Looking specifically at indigenous peoples in the United States, Grosse and Mark, (2023) describe a process where indigenous peoples become beholden to government agencies and private investors for economic and administrative resources when pursuing renewable energy projects. This reduces native peoples' sovereignty over their own data, or as the authors describe,

> when non-Native investors finance the majority of the project, they control the project's data and can take the data if the project fails…. For Native individuals, poverty often prevents the leasing of distributed energy from tribal utilities and implementation of microgrid battery solutions. This barrier persists across communities of color, who hold a larger proportion of rooftop solar potential than majority white communities, but are often unable to pay for high upfront solar costs. (Grosse and Mark 2023, p. 7).

This example not only demonstrates how poverty and socio-economic insecurity compound the power dynamics between energy consumers and other actors in the energy transition, it also illustrates how energy data is embedded in a complex assemblage of technologies, administrative processes, and cultures which determine households' level of understanding and control over their own energy data. The irony is profound: without the option of opting-out, the most socio-economically marginalized are likely to have their energy data harvested even as their status as "hard to reach, hard to serve" is used as a justification to neglect their energy data needs (Bouzarovski and Tirado Herrero 2017; Thompson and Bazilian 2014).

Beyond consent, the continued exclusion of socio-economically and politically marginalized communities in the design of data collection, storage, and sharing processes contributes to social injustices (Hummel et al. 2021). This same pattern can be observed when it comes to energy data. The marginalization and exclusion experienced by Black, Indigenous, and People of Color in the energy transition (Kime et al. 2023; Ramirez et al. 2022) is, in many ways, sealed in through energy data injustices that ignore their rights to privacy and fail to gather information which can be of specific use for them when pursuing energy security.

4.3.2 Conclusion: Addressing Energy Data Inequity

This chapter centers around a process and technology through which energy data gains meaning: the datafication of consumption information through smart meters. The discourse around data justice has emerged in response to a growing awareness of the negative consequences of state-corporate surveillance for everyday people (Bigo et al. 2019; Taylor 2017). Energy data, with its growing volume, velocity, variety, and especially value[1], is being drawn into these debates and critiques. As smart meters and data-driven utilities proliferate, society must grapple with the difficulty in finding the right the balance between leveraging energy data for sustainability and innovation and safeguarding the privacy and agency of all consumers—especially the most vulnerable. The rapid development and deployment of this technology make responsible innovation an important prerogative for the further enlargement of the technology.

The data collection practices of smart meters pose socially and politically salient privacy problems. In addition to potential concerns around personally identifiable information (PII), the forcefulness of data collection by many national and regional governments is questionable. Many places (e.g., France, Norway) do not offer an opt out option to smart meter installation, which effectively makes the installation mandatory (Council of European Energy Regulators 2019; The Connexion 2019); consequently, energy consumption data is being collected without individual or household level consent. This enforcement neglects privacy concerns from the general public. In some of the places that offer an opt out option, there are financial penalties (e.g., California) because the utilities must then send out employees to check on the electricity use for billing purposes (Hess 2014; Lee and Hess 2021). This discriminates against lower income households that cannot afford to opt out and are thus more likely to install smart meters. This ultimately leads to inequitable data collection processes and outcomes in which financially marginalized communities are more likely to have their PII extracted and exploited.

Although not given as much attention as privacy, energy data production also stimulates new concerns related to sustainability. Energy data from smart meters enables both direct and indirect energy savings by providing consumers and utilities with detailed, real-time insights into energy consumption patterns. When users are given accessible feedback on their usage, such as through in-home displays, apps, or reports, they are empowered to modify inefficient behaviors, automate energy management, and participate in demand response programs. On a broader scale, utilities leverage this granular data to optimize grid operations and design targeted incentives, promoting conservation and reduced peak demand.

However, these energy-saving benefits may be partly offset by the rebound effect—a phenomenon where efficiency gains inadvertently lead to increased overall consumption. For example, after achieving savings, individuals may feel justified in using more energy elsewhere, or lower off-peak prices may incentivize greater use during those times, counteracting the initial reductions. Additionally, confidence in smart automation can encourage a ‘set-and-forget’ mentality, where other conservation habits lapse, further diluting net energy savings (Brännlund et al. 2007; Turner 2013). At the system level, increased energy efficiency driven by smart metering may lower overall costs, prompting greater appliance ownership or higher comfort standards across society, which erodes aggregate energy reductions (Azevedo 2014). Thus, while smart meters and the data they generate are crucial tools for advancing energy efficiency and supporting sustainable grid management, their full potential is

[1] See for example the discourse around data being the new oil (Calzada 2021).

best realized when rebound effects are anticipated and mitigated through thoughtful program design and continuous consumer engagement.

To fully engage with energy data production and the many meanings energy data embodies in this process, this chapter focuses on the actors whose consumption is the source of energy data: households. It examines their control over energy data, or lack thereof, and looks at how the level of consumer control over energy data can vary with the socio-economic status of the consumer. It situates the actor in relation to other actors (e.g., energy producers) and to the system at large to examine problem-shifting dynamics affecting energy consumers. By considering energy technologies and energy data itself as actors in the system, we are able to reflect on how technologies, especially a complex of technologies, can act to confine the agency of other system actors, particularly energy consumers.

This chapter also assesses the level of responsibility applied to energy data by considering energy data as an actor that—through its own self-replication—steers our imagination around who should be responsible for it, and in what ways. In the resulting system of interactions, no one is responsible for energy data, very few benefit from it, and some actors—those whose choices produce it themselves are at the losing end of a relationship through and with energy data. While this chapter on energy data production describes the state of play, the book argues that another energy data system is possible. And the need is urgent because it must be acknowledged that the current data system needs to also recognizes the right of the energy poor and the socio-economically marginalized to be counted, considered and prioritized in energy transitions policy (Bouzarovski and Tirado Herrero 2017; Nguyen et al. 2019; Streimikiene et al. 2021).

The question of access and affordability is crucial given the statistics on energy poverty. In a recent report published by the UNDP[2], the authors report that "at least 1.18 billion are energy poor and unable to use electricity, a total that is 60% higher than the 733 million people who lack any electricity connection at all in 2020, according to official data." Looking beyond these numbers are the critical health and educational implications of the underconsumption of energy. Therefore, there is a need for alternatives rooted in responsible innovation principles—particularly of inclusiveness, reflexivity, and responsiveness—are argued to offer possibilities which may allow for the productive use of energy data that also places greater agency and decision-making power in the hands of the energy consumer. In response to this challenge, Chapter 7 describes how data donation and consumption corridors can serve as novel energy data paradigms.

[2] See in https://data.undp.org/blog/1-18-billion-around-the-world-in-energy-poverty.

References

Abele E, Panten N, Menz B (2015) Data collection for energy monitoring purposes and energy control of production machines. Procedia CIRP 29:299–304. https://doi.org/10.1016/j.procir.2015.01.035

Armoogum S, Bassoo V (2019) Privacy of energy consumption data of a household in a smart grid. In: Smart power distribution systems. Elsevier, pp 163–177. https://doi.org/10.1016/B978-0-12-812154-2.00008-0

Azevedo IML (2014) Consumer end-use energy efficiency and rebound effects. Annu Rev Environ Resour 39:393–418. https://doi.org/10.1146/annurev-environ-021913-153558

Bigo D, Isin EF, Ruppert ES (eds) (2019) Data politics: worlds, subjects, rights, Routledge studies in international political sociology. Routledge, Taylor & Francis Group, London , New York

Bouzarovski S, Tirado Herrero S (2017) The energy divide: integrating energy transitions, regional inequalities and poverty trends in the European Union. Europ Urban Reg Stud 24:69–86. https://doi.org/10.1177/0969776415596449

Brännlund R, Ghalwash T, Nordström J (2007) Increased energy efficiency and the rebound effect: effects on consumption and emissions. Energy Econ 29:1–17. https://doi.org/10.1016/j.eneco.2005.09.003

Burget M, Bardone E, Pedaste M (2017) Definitions and conceptual dimensions of responsible research and innovation: a literature review. Sci Eng Ethics 23:1–19. https://doi.org/10.1007/s11948-016-9782-1

Burton J, Kemp D, Barnes R, Parmenter J (2024) Mapping critical minerals projects and their intersection with Indigenous peoples' land rights in Australia. Energy Res Soc Sci 113:103556. https://doi.org/10.1016/j.erss.2024.103556

Calzada I (2021) Data co-operatives through data sovereignty. Smart Cities 4:1158–1172. https://doi.org/10.3390/smartcities4030062

Campos I, Marín-González E (2020) People in transitions: energy citizenship, prosumerism and social movements in Europe. Energy Res Soc Sci 69:101718

Ceci M, Corizzo R, Fumarola F, Ianni M, Malerba D, Maria G, Masciari E, Oliverio M, Rashkovska A (2014) Big data techniques for supporting accurate predictions of energy production from renewable sources. In: Proceedings of the 19th International database engineering & applications symposium on—IDEAS '15. Presented at the the 19th International database engineering & applications symposium. ACM Press, Yokohama, Japan, pp 62–71. https://doi.org/10.1145/2790755.2790762

Council of European Energy Regulators (2019) Implementing technology that benefits consumers int eh clean energy for all european package: selected case studies

Cranmer Z, Steinfield L, Miranda J, Stohler T (2023) Energy distributive injustices: assessing the demographics of communities surrounding renewable and fossil fuel power plants in the United States. Energy Res Soc Sci 100:103050. https://doi.org/10.1016/j.erss.2023.103050

Dayarathna M, Wen Y, Fan R (2015) Data center energy consumption modeling: a survey. IEEE Commun Surv Tutor 18,(1):732–794

Dias Pereira L, Raimondo D, Corgnati SP, Gameiro Da Silva M (2014) Energy consumption in schools—a review paper. Renew Sustain Energy Rev 40:911–922. https://doi.org/10.1016/j.rser.2014.08.010

Eurostat (2025) Energy consumption in households

Forbush JR (2012) Regulating the use and sharing of energy consumption data: assessing California'S SB 1476 smart meter privacy statute. Albany Law Rev 75

Foucault M (2020) Power/knowledge. In: Seidman S, Alexander JC (eds) The New Social Theory Reader. Routledge, pp 73–79. https://doi.org/10.4324/9781003060963-10

Grosse C, Mark B (2023) Does renewable electricity promote Indigenous sovereignty? Reviewing support, barriers, and recommendations for solar and wind energy development on Native lands in the United States. Energy Res Soc Sci 104:103243. https://doi.org/10.1016/j.erss.2023.103243

Hess DJ (2014) Smart meters and public acceptance: comparative analysis and governance implications. Health Risk Soc 16:243–258

Houngbonon GV, Le Quentrec E, Rubrichi S (2021) Access to electricity and digital inclusion: evidence from mobile call detail records. Humanit Soc Sci Commun 8:170. https://doi.org/10.1057/s41599-021-00848-0

Hummel P, Braun M, Tretter M, Dabrock P (2021) Data sovereignty: a review. Big Data Soc 8:2053951720982012. https://doi.org/10.1177/2053951720982012

Jansma SR, Long LAN, Lee D (2023) Understanding energy citizenship: how cultural capital shapes the energy transition. Energies 16:2106. https://doi.org/10.3390/en16052106

Jasanoff S (ed) (2010) States of knowledge: the co-production of science and social order, transferred to digital print. ed, International library of sociology. Routledge, London

Zheng J, Gao DW, Lin L (2013) Smart meters in smart grid: an overview. In: 2013 IEEE green technologies conference (GreenTech). Presented at the 2013 IEEE green technologies conference (GreenTech 2013). IEEE, Denver, CO, pp 57–64. https://doi.org/10.1109/GreenTech.2013.17

Kara S, Bogdanski G, Li W (2011) Electricity metering and monitoring in manufacturing systems. In: Hesselbach J, Herrmann C (eds.) Glocalized solutions for sustainability in manufacturing. Springer Berlin Heidelberg, Berlin, Heidelberg, pp 1–10. https://doi.org/10.1007/978-3-642-19692-8_1

Kime S, Jacome V, Pellow D, Deshmukh R (2023) Evaluating equity and justice in low-carbon energy transitions. Environ Res Lett 18:123003. https://doi.org/10.1088/1748-9326/ad08f8

Koseleva N, Ropaite G (2017) Big data in building energy efficiency: understanding of big data and main challenges. Procedia Eng 172:544–549. https://doi.org/10.1016/j.proeng.2017.02.064

Kurz HD, Schütz M, Strohmaier R, Zilian SS (2022) The Routledge handbook of smart technologies: an economic and social perspective, 1st edn. Routledge, London. https://doi.org/10.4324/9780429351921

Latour B (2005) Reassembling the social: an introduction to actor-network-theory, Clarendon lectures in management studies. Oxford University Press, Oxford , New York

Lee D, Hess DJ (2021) Data privacy and residential smart meters: comparative analysis and harmonization potential. Utilities Policy 70:101188

Lee JT, Freitas J, Ferrall IL, Kammen DM, Brewer E, Callaway DS (2019) Review and perspectives on data sharing and privacy in expanding electricity access. Proc IEEE 107:1803–1819. https://doi.org/10.1109/JPROC.2019.2919306

Lohmöller J, Pennekamp J, Matzutt R, Schneider CV, Vlad E, Trautwein C, Wehrle K (2024) The unresolved need for dependable guarantees on security, sovereignty, and trust in data ecosystems. Data Knowl Eng 151:102301. https://doi.org/10.1016/j.datak.2024.102301

Moghaddam MP, Abdollahi A, Rashidinejad M (2011) Flexible demand response programs modeling in competitive electricity markets. Appl Energy 88:3257–3269. https://doi.org/10.1016/j.apenergy.2011.02.039

Mousavi-Avval SH, Rafiee S, Jafari A, Mohammadi A (2011) Optimization of energy consumption for soybean production using Data Envelopment Analysis (DEA) approach. Appl Energy 88:3765–3772. https://doi.org/10.1016/j.apenergy.2011.04.021

Nguyen TT, Nguyen T-T, Hoang V-N, Wilson C, Managi S (2019) Energy transition, poverty and inequality in Vietnam. Energy Policy 132:536–548. https://doi.org/10.1016/j.enpol.2019.06.001

Ramírez C (2021) Without environmental justice, the renewable energy transition will leave low-income and BIPOC communities behind. JSPG 18. https://doi.org/10.38126/JSPG180306

Ramirez J, Angelino Velázquez D, Vélez-Zapata C (2022) The potential role of peace, justice, and strong institutions in Colombia's areas of limited statehood for energy diversification towards governance in energy democracy. Energy Policy 168:113135. https://doi.org/10.1016/j.enpol.2022.113135

Rowland NJ, Passoth J-H, Kinney AB (2011) Latour's greatest hits, reassembled: review of Bruno Latour's reassembling the social: an introduction to actor-network-theory. SponGe 5:95–99. https://doi.org/10.4245/sponge.v5i1.14968

Depuru SSSR, Wang L, Devabhaktuni V, Gudi N (2011) Smart meters for power grid—challenges, issues, advantages and status. In: 2011 IEEE/PES power systems conference and exposition. Presented at the 2011 IEEE/PES power systems conference and exposition, pp 1–7. https://doi.org/10.1109/PSCE.2011.5772451

Shrouf F, Miragliotta G (2015) Energy management based on Internet of Things: practices and framework for adoption in production management. J Clean Prod 100:235–246. https://doi.org/10.1016/j.jclepro.2015.03.055

Sovacool BK, Hess DJ, Cantoni R, Lee D, Claire Brisbois M, Jakob Walnum H, Freng Dale R, Johnsen Rygg B, Korsnes M, Goswami A, Kedia S, Goel S (2022) Conflicted transitions: exploring the actors, tactics, and outcomes of social opposition against energy infrastructure. Glob Environ Chang 73:102473. https://doi.org/10.1016/j.gloenvcha.2022.102473

Sovacool BK, Kivimaa P, Hielscher S, Jenkins K (2017) Vulnerability and resistance in the United Kingdom's smart meter transition. Energy Policy 109:767–781. https://doi.org/10.1016/j.enpol.2017.07.037

Streimikiene D, Kyriakopoulos GL, Lekavicius V, Siksnelyte-Butkiene I (2021) Energy poverty and low carbon just energy transition: comparative study in Lithuania and Greece. Soc Indic Res 158:319–371. https://doi.org/10.1007/s11205-021-02685-9

Tao S, Ru MY, Du W, Zhu X, Zhong QR, Li BG, Shen GF, Pan XL, Meng WJ, Chen YL, Shen HZ, Lin N, Su S, Zhuo SJ, Huang TB, Xu Y, Yun X, Liu JF, Wang XL, Liu WX, Cheng HF, Zhu DQ (2018) Quantifying the rural residential energy transition in China from 1992 to 2012 through a representative national survey. Nat Energy 3:567–573. https://doi.org/10.1038/s41560-018-0158-4

Taylor L (2017) What is data justice? The case for connecting digital rights and freedoms globally. Big Data Soc 4:205395171773633. https://doi.org/10.1177/2053951717736335

Thapar S (2020) Energy consumption behavior: a data-based analysis of urban Indian households. Energy Policy 143:111571. https://doi.org/10.1016/j.enpol.2020.111571

The Connexion (2019) French court rules against Linky for health reasons [WWW Document]. https://www.connexionfrance.com, https://www.connexionfrance.com/article/French-news/French-court-rules-against-Linky-smart-meters-for-health-reasons-in-Tours. Accessed 24 Sept 2023

Thompson G, Bazilian M (2014) Democratization, energy poverty, and the pursuit of symmetry. Global Pol 5:127–131

Turner K (2013) "Rebound" effects from increased energy efficiency: a time to pause and reflect. Energy J 34:25–42. https://doi.org/10.5547/01956574.34.4.2

US Energy Information Administration (2024) How much energy is consumed in U.S. buildings?

US Energy Information Administration (2023) Electricity consumption in U.S. homes varies by region and type of home

Valverde M (2007) Reassembling the social: an introduction to actor-network theory. By Bruno Latour. Oxford, United Kingdom: Oxford University Press, 2005. pp. 316. $39.95 cloth. Law Soc Rev 41:744–746. https://doi.org/10.1111/j.1540-5893.2007.00319_3.x

Wang J, Gao F, Zhou Y, Guo Q, Tan C-W, Song J, Wang Y (2023) Data sharing in energy systems. Adv Appl Energy 10:100132. https://doi.org/10.1016/j.adapen.2023.100132

Weckert J (ed) (2017) Computer ethics, 1st edn. Routledge. https://doi.org/10.4324/9781315259697

Williams S, Short M (2020) Electricity demand forecasting for decentralised energy management. Energy Built Environ 1:178–186. https://doi.org/10.1016/j.enbenv.2020.01.001

Wu T, Xu D-L, Yang J-B (2021) Decentralised energy and its performance assessment models. Front Eng Manag 8:183–198. https://doi.org/10.1007/s42524-020-0148-7

Zafar R, Mahmood A, Razzaq S, Ali W, Naeem U, Shehzad K (2018) Prosumer based energy management and sharing in smart grid. Renew Sustain Energy Rev 82:1675–1684

Zhang Y, Ma S, Yang H, Lv J, Liu Y (2018) A big data driven analytical framework for energy-intensive manufacturing industries. J Clean Prod 197:57–72. https://doi.org/10.1016/j.jclepro.2018.06.170

Zheng J, Dang Y, Assad U (2024) Household energy consumption, energy efficiency, and household income–evidence from China. Appl Energy 353:122074. https://doi.org/10.1016/j.apenergy.2023.122074

Zhou K, Yang S (2016) Understanding household energy consumption behavior: the contribution of energy big data analytics. Renew Sustain Energy Rev 56:810–819. https://doi.org/10.1016/j.rser.2015.12.001

Zou B, Luo B (2019) Rural household energy consumption characteristics and determinants in China. Energy 182:814–823. https://doi.org/10.1016/j.energy.2019.06.048

Zuboff S (2020) The age of surveillance capitalism: the fight for a human future at the new frontier of power, first trade paperback, edition. PublicAffairs, New York, NY

Zuboff S (2015) Big other: surveillance capitalism and the prospects of an information civilization. J Inf Technol 30:75–89. https://doi.org/10.1057/jit.2015.5

Chapter 5
Energy Data Sharing

5.1 Energy Data as a Store of Value

Energy data has become an invaluable asset in our contemporary energy landscape, serving a multitude of purposes that extend far beyond simple consumption tracking. For instance, the transition towards renewable energy requires a significant increase in energy flexibility and efficiency; both of which require secure, timely, and continuous energy data sharing. Smart grids are the main conduit for data production and sharing in these decentralized energy systems.

Anthi et al. (2018, p. 1) describe smart grids as “a network of intelligent entities that are capable of bidirectional communication and can autonomously operate and interact with each other to deliver power to the end users.” Smart grids not only capture and transmit data on real-time electricity usage, they also capture granular patterns of energy usage, appliance performance, and energy users’ behaviors. The collection of intelligent electricity technologies that we call the smart grid orchestrates and supports various system functions including generation, transmission, distribution, transformation, scheduling and consumption, and enable decentralization and greater integration of alternative power resources (Zheng et al. 2013; Zhou and Yang 2016). In doing so, they generate and share a diverse and complex array of information—or energy data.

Energy data is often positioned in the literature as an object that contributes to the development of solutions to pressing social and economic issues. Viewed as an object, the distribution and sharing of energy data can be justified in terms of the functions that energy data fulfills. Energy data is a key enabler of the (renewable) energy transition, which is organized around three social goals: energy security, environmental sustainability, and economic affordability (i.e., energy trilemma, International Energy Agency (2024)). For instance, energy data may help increase affordability by highlighting patterns of consumption that indicate underserved or vulnerable populations who may be targeted for subsidies. However, energy data cannot fully capture a complex concept like energy poverty, as more nuanced energy poverty indicators

L. A. Nguyen Long and D. Lee, *Energy Data*, Lecture Notes in Energy 46,
https://doi.org/10.1007/978-3-032-19171-7_5

demonstrate (Nussbaumer et al. 2012; Thomson et al. 2017). Moreover, as it often is in with all emerging technologies, tensions often arise between the social goals targeted by energy data (Lee and Nguyen Long 2025). Additionally, and as discussed earlier, efforts to pursue energy security, sustainability, and affordability via energy data may run counter to other core societal values and goals such as privacy and fairness.

5.2 The Plurality of Values Stored in Energy Data

Energy data holds significant social and economic value (Huff and Lee 2020). There are general features which can determine the value of energy data. Wang et al. (2023) classify these features in terms of the form (e.g., accessibility, consistency, comprehensibility, timeliness, security and completeness) and content (e.g., accuracy, scarcity (exclusivity), authenticity, correctness, objectivity, multi-dimensionality, validity reliability, and order of magnitude) of data. They further argue that data that incur lower legal risks and restrictions and face lower moral resistance hold more value (Wang et al. 2023). It is not just the data that is diverse, there is also variation in the ability of different energy stakeholders to derive value from these data. Energy data holds value to the extent that it helps a focal stakeholder meet their specific needs (Wang et al. 2023).

The first type of actor we consider are service providers. Service providers can be micro- or macro-utilities. Although micro-utilities include private entities involved in local energy generation, e.g., community energy or energy hubs, macro-utilities describe more conventional, large-scale energy producers (Lee et al. 2019). Both macro and micro-utilities can employ energy data to optimize grid management, integrate renewable sources, and improve forecasting accuracy. These stakeholders can make use of operational energy data (e.g. voltage levels, transformer loads, and power quality indicators) to avoid localized outages and increase energy reliability.

Service providers can also use energy data to work with third-party service providers who can use the information to improve these companies' services. For instance, the consumption patterns revealed by energy data generated and shared with utility companies through the smart grid enable dynamic pricing, or the ability to charge higher rates during periods of peak consumption (Forbush 2012). The big data analytics on energy data can also be used to generate personalized energy-saving recommendations, for instance by combining energy data with environmental data, such as weather conditions and renewable energy production. In this way, service providers can increase the value that energy data offers end users.

End users are households or businesses that purchase and consume energy (Lee et al. 2019). End users can use data shared through the grid to optimize their energy consumption schedules, for instance by using pricing information to employ home appliances like washing machines, air conditioners and vacuum cleaners when the cost of electricity falls below a certain threshold (Forbush 2012) and they can automate this process by leveraging smart meter technology. Altogether, such behavior change—a practice referred to as "shedding"—can reduce the strain on energy systems and improve grid reliability. The benefits of end user behavior change accrue to end users and the entire system by increasing energy efficiency. Up until now, the energy data consequences that we have discussed can be associated with economic value, primarily, and then with sustainability, as a secondary value. However, many other values are at play in the distribution of energy data.

Governments are also important stakeholders of energy data systems. Policymakers leverage energy data to shape sustainability initiatives, craft effective energy policies, and monitor progress towards decarbonization goals. They sometimes undertake these tasks in partnership with researchers, who use energy data to gather and offer insights into technical and social aspects of energy systems, and their development (Lee et al. 2019). At the same time, they need to design policies that encourage and enable the secure transfer of data in an energy information system.

Lawmakers face several challenges when regulating data sharing. First, regulations may become outdated quickly. Technological complexity makes it difficult to craft laws that keep up with rapid innovations in data collection, storage, and transfer. When it comes to energy data, utilities and regions often rely on different technologies and data formats, which can lead to weak standardization, which makes it difficult to create consistent rules for sharing. Additionally, the ambiguity of consent and ownership raises difficult questions: who truly controls the data—individuals, companies, or governments? At present, it remains unclear whether energy data belongs to consumers, utilities, or third-party providers. Here governments are positioned as actors who can safeguard privacy, security, and autonomy. The extent to which these values are prioritised by government in its regulation of energy data varies from society to society.

Finally, there is a challenge of public trust and transparency. Herein lies a crucial tension: citizens must feel their data rights are protected without stifling beneficial uses of data. As such, lawmakers need to balance the tension between privacy and innovation. Lawmakers must further balance the need for innovation—such as enabling renewable energy integration and efficiency programs—against the risks of under-regulation. Smart grid data is important for improving energy efficiency, but it can reveal intimate details about household behavior, raising significant privacy concerns over how much information should be shared and with whom. At the same time, this data represents a cybersecurity risk, as breaches could expose vulnerabilities in critical energy infrastructure. Finally, transparency and consumer trust are essential: individuals must understand how their data is used and protected if smart grids are to function effectively.

5.3 Who Profits from My Energy Data? And How?

Given the array of energy data stakeholders and as energy data increasingly fuels new value chains, questions arise about who profits from this wealth of information and how. Utility companies often benefit by using energy data to reduce operational costs, tailor energy services, and promote new offerings to customers through targeted marketing or demand response programs. Large technology firms and data centers may profit indirectly, leveraging complex rate structures or infrastructure upgrades funded by public money or customer tariffs. Additionally, third-party analytics providers can monetize insights by selling aggregated or anonymized datasets, or by offering subscription-based services that help businesses optimize their energy usage. It can, for instance, be used by advertisers to design and deploy personalized marketing campaigns (Murrill 2013). Given the multifaceted use and commercial exploitation of energy data, an important counterbalance is the individual's right to control the sharing and dissemination of their personal energy information.

Consumer advocates and data security experts raise concerns about granting unfettered access to energy data to utilities, third party service providers, and outside data marketers, each hungry for a new source of customer data. A core concern relates to the economic exploitation of energy data for use beyond energy consumption contexts (Forbush 2012). For instance, Energy data shared on the smart grid can be used by health insurance companies to determine the patterns of use of medical devices by a household, when considering their premium (Murrill 2013).

Information on personal habits and product usage is valuable for targeted advertising. More concerningly, governments might use smart grid data as a surveillance tool under the guise of deploying it to detect illegal activities inside homes while also gathering other data, for instance on political dissidents (Forbush 2012). Additionally, this data is vulnerable to hackers and other malicious actors seeking unauthorized access or misuse. Many consumers lack awareness or digital literacy necessary to exert meaningful control over their data. Moreover, energy data flows through a complex ecosystem with many intermediaries, making tracking and limiting secondary use difficult in practice. This asymmetry often advantages informed or technologically savvy users while leaving marginalized individuals vulnerable to data exploitation.

No system is secure enough. As smart meter and smart grid use diffuse widely, terabytes of energy data (e.g., granular energy consumption of millions of users) are transmitted by smart meters and power controllers. Given the volume of data that travels through these systems, the risk of a data leak is a major concern, whether the leak be caused by a malicious actor, human error, or system loophole (Zhu et al. 2018). What is more, current regulations currently fall short of providing adequate protection for energy end users and their data. To more fully consider the risks that accompany the distribution of energy data for energy transitions, we leverage the analytical lens offered by problem shifting.

5.4 Problem Shifting in Energy Data Sharing

Given the social and economic value embedded in energy data, it is no surprise that there is a rush to develop and deploy technologies that allow different stakeholders to capitalize on this information through data production and sharing. The move to adopt smart grid technology and to use it for energy data generation and sharing is a global movement. This movement has the potential to trigger problem shifting dynamics.

To review, problem shifting occurs when a system degrades as an (un)-intended consequence of efforts to improve the performance of another (Kim and van Asselt 2016). Problem shifting can be temporal (problems deferred to future generations), spatial (problems shifted to other places), domain driven (problems shifted to other domains), or a combination of any number of these dimensions. Problem shifting may manifest—this chapter argues—as compromises with regards to one social goal, e.g., privacy, in service of another, e.g., sustainability. This lens can be used to reveal how interdependencies lead to tradeoffs between social goals that ultimately benefit energy companies and private intermediaries at cost to energy consumers and other constituencies. The implication is that energy data must be carefully governed.

Problem shifting in energy data production and sharing arises when solving one challenge unintentionally creates new ones. For example, increasing data collection from smart grids can improve efficiency, support renewable integration, and empower consumers with better usage insights. However, this shift may introduce privacy risks, cybersecurity vulnerabilities, and questions over data ownership. Similarly, regulations designed to protect consumer privacy might restrict data availability. To avoid this and to avoid hindering innovation in demand response or energy efficiency programs regulations may allocate burdens of protecting consumer privacy away from technology developers and utilities towards end users. In this way, attempts to optimize one aspect of energy data use—such as reliability or transparency—can shift problems toward other domains, requiring lawmakers to balance competing priorities carefully.

At the same time, problems are shifted from one set of system actors (and domain) to another. The risks related to energy data disproportionately fall on energy end-users, whose privacy is at risk with the use of smart grid systems. Although they are the most vulnerable to privacy breaches, they also command the least control over energy data. Thus, their privacy and autonomy may be shortchanged in favor of short-term economic gains for service providers or the desire to pursue efficiency and flexibility through greater technological surveillance of consumption patterns. The risk of abuse is high, especially when the development and use of smart grid technologies is not inclusive and responsive to the needs of end users, particularly socially and economically vulnerable end users.

5.5 Responsible Innovation: Reconsidering the Security Challenges of Smart Grids

Data sharing is not without its challenges. Overcoming these challenges is critical to reaping the benefits from energy data that have been outlined above. Wang et al. (2023) identify the four core data sharing challenges arise from: (1) the massive volume of data collected and made available through smart grids, which may be beyond the data processing, storage, and management capabilities of senders and receivers of energy data, (2) besides the volume, the complexity of the data itself, and its granular nature makes it difficult to handle; moreover (3) not all the data generating in the grid is of value, and sifting, cleaning, and sorting energy data is both a resource and energy intensive task; finally, (4) handling energy data comes with its own technological requirements, including high processing speed requirements. Beyond these practical challenges, energy data sharing also faces considerable security and privacy challenges, which arise out of the use of smart grids for data collection and transmission.

5.5.1 Security

Security risks emerge when considering data management and transparency. Managing these pluriform data systems is complicated by their decentralized nature and the high levels of interoperability among the devices within the smart grid. This raises the possibility of system breaches and poor data stewardship by utilities (e.g., error handling, weak protocols, and improper session management). Moreover, there is a risk that data is misused for non-energy purposes, such as insurance profiling or surveillance and the use of energy profiles for targeted marketing. At worst, poorly secured energy data could even provide cues about when a building is unoccupied, opening doors to criminal activity.

Digitalization increases the points of vulnerability (or attack surface) to unauthorized access, (accidental) data breaches, theft, and manipulation of energy consumption data (Zhu et al. 2018). The massive volume of consumer data moving through a smart grid is stored at many different locations in the grid and the physical world, each of which is a point of cyber-physical vulnerability (Murrill 2013). Vulnerability is heightened when data is on the move on a wireless network. For instance, the widespread deployment of smart meters provides a high number of targets for malicious hackers, and it is nearly impossible for a utility company to ensure that no smart meter is compromised because of the disaggregated nature of these attacks (Nambundo et al. 2025). Moreover, the high levels of interconnectivity in smart grids implies that a compromised smart meter can rapidly contaminate an entire smart grid system. The resulting security risks, increase system vulnerabilities to malicious cyber-physical attacks and raise significant concerns about the gaps, inconsistencies and inadequacies in current data regulations (McKenna et al. 2012).

This highlights the need to anticipate and prevent energy data security breaches. Cyber-physical attacks can affect end-users in different ways. Utilities risk disruptions and poor decision-making. For instance, a hacker may manipulate information being generated through the monitor to alter electricity bills (Armoogum and Bassoo 2019). Alternatively, spoofing describes a modus operandi where attackers compromise the effectiveness of an energy management system by exploiting fake identities to extract more energy or energy data from the grid and reduce system latency.

Denial of Service (DoS) attacks debilitate or eliminate the network's communications system reducing control over energy flow on the energy management system, while distributed DoS (DDoS) are direct attacks on an energy management system (Anthi et al. 2018). The resulting disruption services can be particularly harmful during periods of extreme weather (e.g. during the height of summer or winter), when power outages can lead to the loss of life and to significant property damage (Armoogum and Bassoo 2019). Accordingly, utility companies are exploring and deploying safeguards against data security risks, including data anonymization, masking techniques, encryption, and data aggregation (Dhanesh Menon et al. 2019; Forbush 2012; Shuaib et al. 2015). However, these privacy safeguards cannot fully safeguard energy data (McKenna et al. 2012; Murrill 2013).

Given the risks summarized above and the weak legal protections given to energy consumers, many researchers have raised concerns about people's ability to control how their data is shared and used by third parties. Although personal data originates from individuals, it often moves between various companies that provide services to these individuals and create value and can be monetized in the process. This creates a mismatch between those who are responsible for the data and those who have an interest in it, which can hinder the smooth flow of personal data among different entities. While it is true that utilities tend to follow a more or less top-down organizational structure, where the roles and responsibilities of different players (e.g. suppliers, consumers, third party service providers) are clearly demarcated and defined, the pluriform uses of energy data and the wide range of stakeholder interests in energy data complicate the task of specifying who should have access to data, and under what circumstances.

A good starting point in this regard would be to consider what constitutes a "right" when it comes to energy data. Wang et al. (2023) identify two types of energy data rights. First are personality rights, which give users control over the commercialization of their own data and allow them to restrain the representation of their personality in the public sphere. An example of personality rights related to energy data is the protection of personal energy consumption information collected via smart meters. Personality rights would allow the individual to control who can access, use, or disclose their energy usage data, ensuring their privacy and preventing misuse—such as unauthorized profiling or surveillance based on their energy consumption patterns (Van Der Sloot 2015). Second, property rights refer to the legal rights individuals or entities hold over energy data as an asset. This includes rights concerning the control, use, access, and disposition of data generated in the energy sector, such as operational data from energy equipment or sensor data collected during energy production or consumption. Although energy data is not covered by copyright, managed through contractual agreements that define who can access and use the data and under what conditions. For instance, a company operating a wind farm may own the meteorological data collected from sensors on its site. Although the company has property rights over this raw data, it often shares this data with partners or contractors under specific contractual terms that govern how the data can be used or further processed.

Determining who has rightful access to and responsibility for energy data is rarely clearcut. Moving beyond legal rights, Ribot and Peluso define "access" as an agent's ability to draw benefits from the property in question, which in our case is energy data. This offers a broader understanding of access which emphasizes the various social forces (social, economic, and political powers) which enable an actor to draw benefit from (or exploit) the focal property (Ribot and Peluso 2003). Access to energy data is crucial because, although end-users maintain control over their personal data kept private, they lose the ability to manage or limit access once other parties collect it. Conversely, service providers hold the power to regulate and restrict other stakeholders' access to both their own and their customers' data (Lee et al. 2019). This highlights the importance of the right of portability. The right to data portability enables individuals to retrieve and reuse the personal data they have provided to different data controllers for their own purposes. This right allows them to safely and efficiently move, copy, or transfer their personal data from one service provider to another, helping them choose the service that best meets their needs (Wang et al. 2023).

5.5.2 *Privacy*

Privacy is a central value driving many contemporary debates and critiques concerning energy data. The most commonly discussed problem associated with energy data storage is the issue around privacy (Asghar et al. 2017; Brown and Kennedy 2017; Hess and Coley 2014; Lee and Hess 2021; McKenna 2018; Miglani et al. 2020; Sovacool et al. 2017; Teng et al. 2022; Zethmayr and Kolata 2018). Because energy data is collected frequently and over a long period of time, the data can be used to identify or infer household appliance use (Kaatz 2017), socio-economic status (Beckel et al. 2014), and security issues such as what people are doing in their home and if they are away from home (McDaniel and McLaughlin 2009).

Some more sensitive personal information has been identified using energy data. Schirmer et al. (2021) showed that energy data could be used to identify the multimedia content played by a TV or a monitor device. Studies have shown that other sensitive data such as religion and health (Anderson and Fuloria 2010; Erkin et al. 2013) can also be detected using energy data. Consequently, governments have introduced regulations and practices to protect personally identifying information (PII) and other personal information through various measures. In sum, the detailed nature of smart meter data introduces new privacy threats. The smart grid raises privacy and security concerns due to their real-time tracking, data transmission, and remote-control capabilities it enables. This sensitive data, if misused, could expose personal habits or be vulnerable to cyberattacks. Legal implications exist under laws governing digital privacy, surveillance, data retention, and consumer rights. In the past, the collection of energy data did not raise privacy concerns because energy data was directly accessed from buildings over longer time periods, and none of the data collected were specific to appliances or devices in the home (Forbush 2012).

Unlike traditional meter readings, smart data may reveal intimate details about occupants' routines, appliance use, and presence or absence in the home. The availability of fine-grained big date on household energy use can reveal highly sensitive information. For instance, reveal where in a home individuals are present and at what time (Depuru et al. 2011). Connecting to the smart grid, a smart meter data can identify which appliances are used and when, because each appliance has a unique electricity load pattern. This reveals occupant behavior, such as presence, number of residents, daily routines, and ownership of devices like alarms or medical equipment, posing privacy risks (Soares et al. 2021).

Consequently, concerns have been raised about the potential for this information to be accessed by unauthorized third parties, including hackers, corporations, or government agencies. Smart meters' detailed data can reveal sensitive consumer information—presence at home, health, habits, and device usage. Such granularity risks misuse if accessed without authorization, including exploitation by outsiders. Altogether, energy data raises concerns about individual privacy, autonomy, anonymity, and the accountability of utility companies to their customers, who may even be unaware that their data is being collected (Forbush 2012).

To have privacy means to have the ability to control and constrain personal information flows, including others' access to this information and how it is shared and used (Hummel et al. 2021). The extent to which privacy is considered and treated as a fundamental human right varies across national contexts (Bygrave 2014). In the European Union, privacy is governed under a complex of strict, comprehensive rules on personal data processing, rights of individuals, cross-border transfers, and heavy fines adopted in the 2018 General Data Protection Regulation (GDPR). The GDPR has had a sizable impact on global digital policy, and inspired a groundswell of privacy regulation across the globe (Cervi 2022; Goddard 2017), influencing privacy laws in California (USA), Canada, and South Africa (Wang et al. 2023). When it comes to energy data privacy in the E.U., another relevant law is the 2022 EU Directive on a common level of cybersecurity in the Union (NIS2 Directive[1]), which targets critical infrastructure sectors like energy.

South Korea's PIPA (Personal Information Protection Act), last amended in 2023, is considered one of the strictest privacy regulations in Asia (Greenleaf and Park 2014; Ko et al. 2017; Lim and Oh 2025), if not the world, and operates alongside the Act on the Use and Protection of Credit Information ("CIA") and the Act on the Protection and Use of Location Information ("LIA") in governing data in the country. In the U.S., privacy is governed through a patchwork of state-level and sectoral federal laws like the California Consumer Privacy Act. At the same time, 1996's Health Insurance Portability and Accountability Act governs the management and sharing of health data. Overall, around 82% of the world's population are living in one of the 144 countries that have adopted privacy laws, with varying levels of stringency among them. 49 remaining countries lack comprehensive privacy regulations. Regulatory frameworks, like the GDPR in Europe, have attempted to address these risks, but enforcement and best practices remain inconsistent worldwide.

[1] https://digital-strategy.ec.europa.eu/en/policies/nis2-directive.

There are also attempts to allay privacy concerns at the point of data production. Restricting the frequency of data sampling is the most obvious and widely spread approach (Rajagopalan et al. 2011). In Europe, smart meters are active in somewhere between 5 and 30 min interval (Nijhuis et al. 2019), although many countries have restricted it to 30 min sampling frequency (Lee and Hess 2021; UK Power Networks 2019). In the U.S., the usual data collection ranges from 15 to 30 min (Lee and Hess 2021; Page et al. 2015). By capping the frequency, regulators and utilities aim to retain sufficient temporal resolution for system optimization while reducing the risk of detailed behavioral profiling.

Another method commonly used is data aggregation. U.S. California established the "15/15 Rule," which states that a utility may share consumption data only when it has been aggregated across a group of at least fifteen customers, with no individual customer accounting for more than fifteen percent of the group's total usage. A similar practice is carried out in Colorado, whereas many ESCOs in Illinois use the "4/80 Rule," which means utilities may aggregate energy data from as few as four customers, provided that no single customer contributes more than 80% of the group's total consumption (Elevate Energy 2013) and New York as a "4/50 Rule" (State of New York Public Service Commission 2018). In Europe, data aggregation is also discussed, particularly in the United Kingdom (Department for Energy Security and Net Zero 2018), although not as frequently regulated through rules as in the U.S.

Nevertheless, advances in data analytics and the integration of smart meters with other digital infrastructures continue to challenge the adequacy of these limits, raising questions about whether existing frequency restrictions and data aggregation are sufficient to safeguard consumer privacy in the long term. Therefore, we must question whether the regulatory logic that underpins them remains fit for purpose. We argue considering the digital footprint that we leave behind, coupled with smart meter data, our privacy is very much at risk. This makes it necessary to revisit existing governance frameworks, ensuring that privacy safeguards keep pace with technological change and with the expanding integration of digital infrastructures into the energy sector. Clearly there is a need for more expansive solutions.

5.6 Conclusions: Addressing Energy Data Vulnerabilities

The reasons for requiring responsibility and liability for energy data security—we hope—is sufficiently established in this chapter. Privacy breaches carry serious consequences when they occur, and the limited agency given to energy consumers in controlling their own energy data remains problematic. This section argues that the accountability should span end to end: from data generation (through smart meters) through storage analysis to deletion and rectification.

Accountability begins with the point of data generation. Smart meters and other digital devices record consumption data at increasingly granular levels, producing information that can reveal not only energy usage but also patterns of behavior, occupancy, and lifestyle. Once collected, energy data are typically transmitted to centralized databases or cloud platforms. Here, an additional dimension of sustainability becomes a concern with the increasing demand for data storage and analysis. Access and sharing of energy data constitute a further area requiring robust accountability. Clear liability is necessary to govern who grants access, under what conditions, and how consent is documented and potentially revoked. In the absence of enforceable standards, consumers risk having their data commodified or misused, often without their knowledge. Additionally, the use and analysis of energy data is important. Algorithms applied to energy data can yield valuable insights for efficiency and grid management but they may also generate discriminatory outcomes or intrusive inferences about household routines. Liability should be invoked where analytic practices undermine consumer privacy, reinforce social inequalities or contribute to exclusionary practices. Finally, deletion, retention, and rectification represent the end of the data life cycle—but remains to be relevant. Consumers should have enforceable rights to request correction of inaccurate records, deletion of obsolete data, and limits on retention periods. Breach notification and compensation mechanisms also fall under this domain, ensuring that when failures occur, affected consumers receive timely communication and meaningful redress.

Two broad domains of energy data call for particularly clear responsibility and liability: privacy and sustainability. First, privacy has been the subject of extensive legal and regulatory attention worldwide. From the European Union's General Data Protection Regulation (GDPR) and the U.K.'s Data Protection Act to South Korea's Personal Information Protection Act and the various state-level rules and voluntary codes in the U.S., governments have developed detailed provisions governing the collection, processing, sharing, and retention of personal energy data. These frameworks recognize that smart meter readings can reveal intimate details of household life, such as occupancy patterns and daily routines, and thus require strict consent procedures, limits on secondary uses, and mechanisms for redress in the event of a breach.

The second domain, which at the present time receives comparatively less scrutiny, concerns the sustainability of energy data, particularly the environmental costs associated with large-scale data storage and processing. The growth of smart grids and the proliferation of high-frequency metering generate vast volumes of data that must be transmitted, stored, and analyzed, often in energy-intensive data centers. This creates a paradox: digital systems intended to enable energy efficiency can themselves become significant sources of energy consumption and carbon emissions. Assigning responsibility and liability for this hidden energy footprint—through requirements for efficient data architectures, limits on unnecessary retention, and transparency about the carbon intensity of data centers—would ensure that the digital infrastructure supporting the energy transition does not undermine its environmental objectives. Together, these two domains underscore the need for a comprehensive governance approach that protects personal privacy while also addressing the material and ecological impacts of the expanding energy data economy.

Achieving a holistic and comprehensive governance approach that does not sacrifice privacy for the energy transition requires a mindset shift with regards to how we perceive energy data. Responsible innovation can help address problem shifting in energy data by ensuring that new technologies and policies are developed with foresight, inclusivity, and adaptability. Using energy data as a boundary object for engaging stakeholders—consumers, utilities, regulators, and technology providers—early in the process, potential trade-offs between privacy, innovation, and security can be identified before they escalate. This engagement can also increase transparency and accountability, which in turn can also build trust, ensuring that data is shared in ways that align with societal values while protecting individual rights.

Participatory and anticipatory governance challenges regulators to foresee unintended consequences, such as how stricter privacy protections might limit energy efficiency gains, and design flexible rules to mitigate them. By means of continuous monitoring and adaptive regulation, policies may evolve alongside technological advances, reducing the risk that solving one problem in energy data creates new, unintended dilemmas elsewhere.

Because responsiveness to privacy institutions and inclusion of energy consumers is lacking in the current energy sharing system, remedying the challenges tied to energy data sharing (i.e., security and privacy) requires greater engagement of energy consumers from different walks of life in technological development and deployment decision-making. However, encouraging and maintaining cooperation between different stakeholders (e.g., end users, service providers, technology providers, grid operators, etc.) can be difficult as these actors have different interests and backgrounds (see Sect. 5.5.2). Finding a way for them to engage collaboratively with one another can be particularly challenging when there is distrust or a history of conflict between some or all of the parties involved (Ansell and Gash 2008). While diversity is often proposed to improve decision-making and design, it can be difficult to manage.

Given the diverse stakeholder interests and valuation of energy data and the potential for societal tensions that may arise in consequence, it may be useful to think of energy data as a boundary object around which diverse communities and stakeholder groups organize and mobilize to achieve and maintain consensus (Leigh Star 2010; Star and Griesemer 1989). Boundary objects are concepts, tools, or artifacts that are flexible enough to be understood differently by various groups but robust enough to maintain a common identity across those groups. The 'interpretive flexibility' of boundary objects enables them to be sufficiently robust to offer a shared understanding (or serve as shared reference points or tools of translation) while having the elasticity to be adapted to local contexts and needs. Energy transitions create opportunities to ascribe new meanings to energy data, allowing them to function well as boundary objects. Boundary objects—like energy data—create and maintain a space for cooperative work by establishing shared understandings among stakeholders with differing interpretations and goals.

Boundary objects help different stakeholders navigate the chaotic process of socio-technical transitions by working together. This applies to the current energy transition, where energy data support the development and deployment of collective constructions such as visions, agendas, or roadmaps. Energy data can be used to articulate and mediate diverse perspectives and rationalities among key actors, such as policymakers, industry players, communities, and researchers, thus enabling transitions governance. Under this view, big data artifacts (e.g., datasets, reports, visualizations) act as boundary objects bridge diverse communities (e.g., utility managers, consumers, and policy makers) and enable collaboration between them by offering a common yet flexible interface for interpreting complex information according to varied local needs.

The application of this concept to energy data implies that some governance modes may be preferential to others in addressing the policy challenges and opportunities that arise from the rapid digitalization of energy data. The focus on cooperation in regulating energy transitions via energy data demands participatory governance that addresses democratic challenges arising from representing diverse, sometimes conflicting, stakeholder interests. This complexity is evident in debates on data sovereignty, security, and surveillance within energy systems. Effective governance must balance these plural interests through inclusive dialogue and decision-making processes that enhance legitimacy and responsiveness in energy policy (Radtke and Renn 2024). Such participatory frameworks are essential to navigate tensions and ensure trust, accountability, and equitable outcomes in managing energy data transitions. Here is where responsible innovation may also come into play.

References

Anderson R, Fuloria S (2010) On the security economics of electricity metering

Ansell C, Gash A (2008) Collaborative governance in theory and practice. J Public Adm Res Theory 18:543–571. https://doi.org/10.1093/jopart/mum032

Anthi E, Javed A, Rana O, Theodorakopoulos G (2018) Secure data sharing and analysis in cloud-based energy management systems. In: Longo A, Zappatore M, Villari M, Rana O, Bruneo D, Ranjan R, Fazio M, Massonet P (eds) Cloud infrastructures, services, and IoT systems for smart cities, Lecture notes of the institute for computer sciences, social informatics and telecommunications engineering. Springer International Publishing, Cham, pp 228–242. https://doi.org/10.1007/978-3-319-67636-4_24

Armoogum S, Bassoo V (2019) Privacy of energy consumption data of a household in a smart grid. In: Smart power distribution systems. Elsevier, pp 163–177. https://doi.org/10.1016/B978-0-12-812154-2.00008-0

Asghar MR, Dán G, Miorandi D, Chlamtac I (2017) Smart meter data privacy: a survey. IEEE Commun Surv Tutor 19:2820–2835. https://doi.org/10.1109/COMST.2017.2720195

Beckel C, Sadamori L, Staake T, Santini S (2014) Revealing household characteristics from smart meter data. Energy 78:397–410. https://doi.org/10.1016/j.energy.2014.10.025

Brown A, Kennedy R (2017) Regulating intersectional activity: privacy and energy efficiency, laws and technology. Int Rev Law, Comput Technol 31:340–369. https://doi.org/10.1080/13600869.2017.1371576

Bygrave LA (2014) Data privacy law: an international perspective. Oxford University Press. https://doi.org/10.1093/acprof:oso/9780199675555.001.0001

Cervi GV (2022) Why and how does the EU rule global digital policy: an empirical analysis of EU regulatory influence in data protection laws. Digit Soc 1:18. https://doi.org/10.1007/s44206-022-00005-3

Department for Energy Security and Net Zero (2018) Smart metering implementation programme: review of the data access and privacy framework

Depuru SSSR, Wang L, Devabhaktuni V, Gudi N (2011) Smart meters for power grid—challenges, issues, advantages and status. In: 2011 IEEE/PES power systems conference and exposition. Presented at the 2011 IEEE/PES power systems conference and exposition, pp 1–7. https://doi.org/10.1109/PSCE.2011.5772451

Dhanesh Menon V, Trilok Kumar J, Sabhanayagan M, Ramkumar A, Rajesh K (2019) Cyber security for smart meters. In: 2019 IEEE international conference on intelligent techniques in control, optimization and signal processing (INCOS). Presented at the 2019 IEEE international conference on intelligent techniques in control, optimization and signal processing (INCOS). IEEE, Tamilnadu, India, pp 1–5. https://doi.org/10.1109/INCOS45849.2019.8951407

Elevate Energy (2013) Aggregated data access: the 15/15 rule in Illinois and beyond.

Erkin Z, Troncoso-pastoriza JR, Lagendijk RL, Perez-Gonzalez F (2013) Privacy-preserving data aggregation in smart metering systems: an overview. IEEE Signal Process Mag 30:75–86. https://doi.org/10.1109/MSP.2012.2228343

Forbush JR (2012) Regulating the use and sharing of energy consumption data: assessing California's SB 1476 smart meter privacy statute. Albany Law Rev 75

Goddard M (2017) The EU general data protection regulation (GDPR): European regulation that has a global impact. Int J Mark Res 59:703–705. https://doi.org/10.2501/IJMR-2017-050

Greenleaf G, Park W (2014) South Korea's innovations in data privacy principles: Asian comparisons. Comput Law Secur Rev 30:492–505. https://doi.org/10.1016/j.clsr.2014.07.011

Hess DJ, Coley JS (2014) Wireless smart meters and public acceptance: the environment, limited choices, and precautionary politics. Public Underst Sci 23:688–702

Huff E, Lee J (2020) Data as a strategic asset: improving results through a systematic data governance framework. In: SPE Latin American and Caribbean petroleum engineering conference. Presented at the SPE Latin American and Caribbean petroleum engineering conference. SPE, Virtual, p D031S013R001. https://doi.org/10.2118/198950-MS

Hummel P, Braun M, Tretter M, Dabrock P (2021) Data sovereignty: a review. Big Data Soc 8:2053951720982012. https://doi.org/10.1177/2053951720982012

International Energy Agency (2024) World energy outlook. https://www.iea.org/reports/world-energy-outlook-2024

Kaatz J (2017) Resolving the conflict between new and old: a comparison of New York, California and other state DER proceedings. Electr J 30:6–13. https://doi.org/10.1016/j.tej.2017.10.005

Kim RE, van Asselt H (2016) Global governance: problem shifting in the Anthropocene and the limits of international law. In Morgera E, Kulovesi K (eds) Research handbook on international law and natural resources. Edward Elgar Publishing. https://doi.org/10.4337/9781783478330.00039

Ko H, Leitner J, Kim E, Jeong J (2017) Structure and enforcement of data privacy law in South Korea. Int Data Priv Law 7:100–114. https://doi.org/10.1093/idpl/ipx004

Lee D, Hess DJ (2021) Data privacy and residential smart meters: comparative analysis and harmonization potential. Util Policy 70:101188

Lee D, Nguyen Long LA (2025) Privacy challenges of automated vehicles: merging contextual integrity and responsible innovation frameworks. Transp Res Interdiscip Perspect 32:101536. https://doi.org/10.1016/j.trip.2025.101536

Lee JT, Freitas J, Ferrall IL, Kammen DM, Brewer E, Callaway DS (2019) Review and perspectives on data sharing and privacy in expanding electricity access. Proc IEEE 107:1803–1819. https://doi.org/10.1109/JPROC.2019.2919306

Leigh Star S (2010) This is not a boundary object: reflections on the origin of a concept. Sci Technol Hum Values 35:601–617. https://doi.org/10.1177/0162243910377624

Lim S, Oh J (2025) Navigating privacy: a global comparative analysis of data protection laws. IET Inf Secur 2025:5536763. https://doi.org/10.1049/ise2/5536763

McDaniel P, McLaughlin S (2009) Security and privacy challenges in the smart grid. IEEE Secur Priv 7:75–77. https://doi.org/10.1109/MSP.2009.76

McKenna R (2018) The double-edged sword of decentralized energy autonomy. Energy Policy 113:747–750. https://doi.org/10.1016/j.enpol.2017.11.033

McKenna E, Richardson I, Thomson M (2012) Smart meter data: balancing consumer privacy concerns with legitimate applications. Energy Policy 41:807–814. https://doi.org/10.1016/j.enpol.2011.11.049

Miglani A, Kumar N, Chamola V, Zeadally S (2020) Blockchain for internet of energy management: review, solutions, and challenges. Comput Commun 151:395–418. https://doi.org/10.1016/j.comcom.2020.01.014

Murrill BJ (2013) Smart meter data: privacy and cybersecurity. Congressional Research Service

Nambundo JM, De Souza Martins Gomes O, De Souza AD, Machado RCS (2025) Cybersecurity and major cyber threats of smart meters: a systematic mapping review. Energies 18:1445. https://doi.org/10.3390/en18061445

Nijhuis M, Vermeltfoort N, Bernards R (2019) Applying smart meter data to low voltage network planning. In: 25th international conference on electricity distribution

Nussbaumer P, Bazilian M, Modi V (2012) Measuring energy poverty: focusing on what matters. Renew Sustain Energy Rev 16:231–243. https://doi.org/10.1016/j.rser.2011.07.150

Page J, McParland C, Piette MA, Czarnecki S (2015) Design of an open smart energy gateway for smart meter data management (No. LBNL—182358, 1248928). Lawrence Berkeley Natinoal Laboratory. https://doi.org/10.2172/1248928

Radtke J, Renn O (2024) Participation in energy transitions: a comparison of policy styles. Energy Res Soc Sci 118:103743. https://doi.org/10.1016/j.erss.2024.103743

Rajagopalan SR, Sankar L, Mohajer S, Poor HV (2011) Smart meter privacy: a utility-privacy framework. In: 2011 IEEE international conference on smart grid communications (SmartGridComm). Presented at the 2011 IEEE international conference on smart grid communications (SmartGridComm), pp 190–195. https://doi.org/10.1109/SmartGridComm.2011.6102315

Ribot JC, Peluso NL (2003) A theory of access. Rural Sociol 68:153–181. https://doi.org/10.1111/j.1549-0831.2003.tb00133.x

Schirmer PA, Mporas I, Sheikh-Akbari A (2021) Identification of TV channel watching from smart meter data using energy disaggregation. Energies 14:2485. https://doi.org/10.3390/en14092485

Shuaib K, Trabelsi Z, Abed-Hafez M, Gaouda A, Alahmad M (2015) Resiliency of smart power meters to common security attacks. Procedia Comput Sci 52:145–152. https://doi.org/10.1016/j.procs.2015.05.049

Soares N, Monteiro P, Duarte FJ, Machado RJ (2021) Extending the scope of reference models for smart factories. Procedia Comput Sci 180:102–111. https://doi.org/10.1016/j.procs.2021.01.134

Sovacool BK, Kivimaa P, Hielscher S, Jenkins K (2017) Vulnerability and resistance in the United Kingdom's smart meter transition. Energy Policy 109:767–781. https://doi.org/10.1016/j.enpol.2017.07.037

Star SL, Griesemer JR (1989) Institutional ecology, 'Translations' and boundary objects: amateurs and professionals in Berkeley's Museum of Vertebrate Zoology, 1907–39. Soc Stud Sci 19:387–420. https://doi.org/10.1177/030631289019003001

State of New York Public Service Commission (2018) Order of adopting whole building energy data aggregation standard

Teng F, Chhachhi S, Ge P, Graham J, Gunduz D (2022) Balancing privacy and access to smart meter data

Thomson H, Bouzarovski S, Snell C (2017) Rethinking the measurement of energy poverty in Europe: a critical analysis of indicators and data. Indoor Built Environ 26:879–901. https://doi.org/10.1177/1420326X17699260

UK Power Networks (2019) Data privacy plan for access to smart meter consumption data

Van Der Sloot B (2015) Privacy as personality right: why the ECtHR's focus on ulterior interests might prove indispensable in the age of "Big Data." Utrecht J Int Eur Law 31:25–50. https://doi.org/10.5334/ujiel.cp

Wang J, Gao F, Zhou Y, Guo Q, Tan C-W, Song J, Wang Y (2023) Data sharing in energy systems. Adv Appl Energy 10:100132. https://doi.org/10.1016/j.adapen.2023.100132

Zethmayr J, Kolata D (2018) The costs and benefits of real-time pricing: an empirical investigation into consumer bills using hourly energy data and prices. Electr J 31:50–57. https://doi.org/10.1016/j.tej.2018.02.006

Zheng J, Gao DW, Lin L (2013) Smart meters in smart grid: an overview. In: 2013 IEEE green technologies conference (GreenTech). Presented at the 2013 IEEE green technologies conference (GreenTech 2013). IEEE, Denver, CO, pp 57–64. https://doi.org/10.1109/GreenTech.2013.17

Zhou K, Yang S (2016) Understanding household energy consumption behavior: the contribution of energy big data analytics. Renew Sustain Energy Rev 56:810–819. https://doi.org/10.1016/j.rser.2015.12.001

Zhu L, Li M, Zhang Z, Du X, Guizani M (2018) Big data mining of users' energy consumption patterns in the wireless smart grid. IEEE Wireless Commun 25:84–89. https://doi.org/10.1109/MWC.2018.1700157

Chapter 6
Energy Data Management and Storage

6.1 How Much Data is Being Saved

UK Power Networks, an energy service company (ESCO) servicing London, the South East and East of England, led Low Carbon London project from November 2011 to February 2014. They sampled 5,567 London households and took meter readings every half hour. This data was published online in 2015 (UK Power Networks 2015) and the 5,567 households that participated in the project comprise of approximately 10 GB.

It is not difficult to extrapolate how much energy data we create from this data. Approximately 65.7 MB of energy data is created by 1,000 households in London every month based on this data. The Office for National Statistics state that in 2023, there were 28.4 million households in the U.K. (United Kingdom Office for National Statistics 2025). Therefore, we can extrapolate that approximately 1.85 TB of energy data produced by the U.K. alone every year. In the U.S., which has 132.2 million households (U.S. Census Bureau 2024), 8.48 TB of data is produced every year.

This may not seem like much—after all, many of us possesses external hard drives with terabytes and some laptops also come with terabytes of drive space. However, three things must be noted before we quickly dismiss the potential significance of these numbers. First, energy data is saved every year and they are rarely deleted, which means we add to the existing data continuously. Using the calculation above, the U.S. alone produces 84.8 TB of energy data every ten years. If we take other countries into account especially with the global trend of mandating or similar strong enforcements of installing smart meters, we expect this number of substantially increase every year.

Second, the above calculation is performed every half hour. Although this is more or less the standard practice globally, some states allow more frequent sampling of energy data especially for commercial and industrial energy data (Lee and Hess 2021). Furthermore, real time pricing, which would allow all the benefits of having smart meter through more easily accessible energy bill, more efficient grid management, and even the possible integration of renewable energy, requires energy data to

L. A. Nguyen Long and D. Lee, *Energy Data*, Lecture Notes in Energy 46,
https://doi.org/10.1007/978-3-032-19171-7_6

be collected every five minutes, not half hour. Consequently, the storage that energy data will require is expected to substantially increase.

Third, the data provided above by UK Power Networks is simple. It has 167 million rows, but it also only has 4 columns (i.e., meter ID, energy pricing tariff, datetime, and kWh/per half hour). Most energy data made available to the public is made anonymous and only shows a small portion of the available data because energy data, particularly those that are taken frequently, pose a severe privacy risk. We can then assume that energy data collected and kept by ESCOs is substantially bigger in size than what we see provided by the UK Power Networks.

In light of these considerations, energy data must be understood not merely as a by-product of smart metering but as a rapidly expanding digital product with material, institutional, and ethical consequences. Its accumulation, frequency, and granularity are shaped by technological standards, policy choices, and market demands, but its long-term maintenance and storage remain largely underexamined. The scale and velocity of energy data production will only accelerate as smart meters become ubiquitous, sampling intervals shorten, and integrated systems demand ever more granular information for optimization and control. This trajectory calls for anticipatory governance that recognizes energy data as a critical component of the energy transition. This carries substantial infrastructural burdens and poses non-trivial risks to privacy, equity, and sustainability. It is no longer sufficient to treat data storage and management as secondary technical concerns; rather, they must be foregrounded in policy design, regulatory frameworks, and infrastructural planning. The challenge before us is not only how to store and manage energy data efficiently, but how to govern it responsibly.

6.2 Energy Data Management

As outlined in Chap. 4, the collection of energy data has become increasingly sophisticated, driven by the widespread deployment of smart meters and the integration of digital infrastructures into energy systems. This chapter shifts focus from collection to management, examining how energy data is stored, maintained, and (not) accessed. Although often treated as a technical or administrative function, energy data management is deeply entangled with regulatory, institutional, and ethical considerations. It reflects broader questions about privacy, sustainability, ownership and control within the digital energy transition sphere. As the volume, frequency, and granularity of energy data expand, so too do the challenges associated with its long-term management. This section critically investigates the current landscape of energy data management, emphasizing the need for anticipatory and reflexive approaches that align technical practices with public accountability, infrastructural sustainability, and data justice.

Once collected by smart meters, energy consumption data is transmitted for processing and storage through centralized systems. The entities authorized to access or manage this data vary by country and regulatory framework. In most cases, energy service companies (ESCOs), utilities, or grid operators oversee data management. However, some countries have established dedicated institutions for this purpose. For example, Norway created Elhub, a national data management platform that processes, stores, and anonymizes energy data to ensure transparency, privacy, and interoperability across market actors (Cegal 2025; Elhub 2025). However, there are few such examples to offer. Before the rollout of smart meters, utility employees would visit individual households to manually record readings from analog meters, typically on a monthly basis. With the adoption of smart metering infrastructure, data is now collected automatically—usually at 30 min intervals for residential customers, though in some contexts, data can be recorded as frequently as every 15 min. This shift not only increases the volume and granularity of available data but also transforms how energy systems are monitored, optimized, and governed (Lee and Hess 2021).

Then these energy data are managed using Meter Data Management Systems (MDMS), which is special software platform or a database with analytical tools that supports the storage, validation, cleaning, organization, and distribution of the vast quantities of energy data (Matheson et al. 2004). These systems constitute the foundational layer of Advanced Metering Infrastructure (AMI), facilitating the transformation of high-frequency, raw consumption data into structured formats that support core utility functions such as billing, demand forecasting, grid analytics, and consumer-facing services.

One of the core functions of MDMS is to perform validation, editing, and estimation (VEE), which ensures incoming data conforms to expected formats and ranges (validation), fills in gaps where there is missing data caused by transmission or device failures using statistical models (estimation), and corrects identifiable errors (edit). Furthermore, MDMS aggregates data into intervals. Using these cleaned meter data MDMS interacts and communicates with billing systems, outage management systems, consumer information system, enterprise resource planning power quality management and load forecasting systems, and others to ensure a smooth running of the distribution of electricity.

Smart meters are capable of storing their own data for periods ranging from seven days to up to three years, depending on the frequency of data collection and device configuration (BituoTechnik 2025; SSE Energy Solutions 2025). However, the duration for which these data are retained ESCOs or data management platforms remains largely opaque. Given that long-term, high-frequency panel data enables more accurate forecasting, system optimization, and customer profiling, it is reasonable to assume that such data are rarely deleted. Instead, they are likely archived and accumulated over time. This extended retention reflects the growing strategic value of energy data, not only for operational efficiency but also for market prediction, policy modeling, and the development of data-intensive technologies within the energy transition.

Several regulatory frameworks, such as the E.U.'s General Data Protection Regulation (GDPR), California's Consumer Privacy Act (CCPA), and South Korea's Personal Information Protection Act (PIPA), have been introduced to limit the collection, retention, and misuse of personally identifiable information (PII), which may include data generated within the energy sector. However, a key limitation shared by these regimes is their ambiguity regarding the classification and treatment of energy data—particularly when such data are anonymized or aggregated. In many cases, anonymized meter data is not considered PII and thus falls outside the scope of these privacy protections, despite growing evidence that re-identification is possible with high-frequency consumption data. To date, there are no publicly documented cases in which these regulations have been directly applied to restrict or sanction the collection of smart meter data by utilities or data management entities. The Federal Commissioner for Data Protection in Germany and the U.K.'s Information Commissioner's Office have published advisory guidelines, but they are non-binding and tend to focus on best practices rather than enforcement (Department for Business, Energy and Industrial Strategy 2021; Federal Office for Information Security 2025). Therefore, this regulatory gap underscores the need for clearer guidance on the status of energy data within broader data protection frameworks.

6.3 Accidents and Non-accidents of Energy Data Storage

Several documented cases illustrate how energy data breaches have compromised sensitive consumer information. A notable instance occurred in 2020, when the People's Energy company, a Scotland based energy supplier, reported a cyberattack on its customer database. The breach resulted in the exposure of approximately 270,000 individuals' PII, including names, addresses, dates of birth, telephone numbers, tariff details, and energy meter identifiers (Kleinman 2020). A similar incident happened in Nova Scotia, Canada in 2025, where Nova Scotia Power had been attacked, which resulted in data breach of almost 300,000 people's names addresses, driver's license numbers, Canadian Social Insurance numbers, bank account details, service requests, customer payment, billing and credit history, and customer correspondence (Greig 2025). Together, these cases underscore the urgent need for accountability and liability discussion of managing, storing, and analyzing energy data.

Some data breaches are difficult to be accounted as "accidents" as they may actually represent carelessness of technology developers and energy suppliers. During the February snowstorm in Texas in 2024, smart meters produced by Landis+Gyr and deployed by major utilities such as Austin Energy and Oncor broadcasted unencrypted uptime information, revealing the precise duration since each device's last power outage. By collecting these signals across a wide geographic area, it became possible to link unique meter identifiers and GPS coordinates to individual households and businesses, effectively mapping which customers retained power during the crisis. This data breach also led to the possibility of inferring which buildings (e.g., hospitals) were prioritized during blackout management. This leak comprised consumer privacy by disclosing highly granular information about energy availability and outage patterns at identifiable premises (Thalen 2021). In short, such exposure show how inadequate security practices in energy infrastructure can transform operational data into a serious privacy vulnerability, eroding public confidence in both utilities and technologies they deploy. These "normal accidents" become inevitable in a system characterized by complexity and weak responsibility (Perrow 2011).

Although it is unlikely that such breaches would lead directly to the unauthorized withdrawal of funds, given the layered security protocols in place with in the banking sector, the compromised data nonetheless creates enduring vulnerabilities. Once personal identifiers such as names, addresses, and phone numbers circulate beyond secure systems, they can be repeatedly exploited for phishing schemes, social engineering, or identify fraud. In this sense, the breach permanently alters the security landscape for affected individuals, as their personal information can no longer be fully retrieved or safeguarded, which leaves them exposed to ongoing risks.

These cyberattacks are not only harmful to consumers but also to companies because compensations are in order when these data breaches happen. People's Energy in Scotland ceased operation in 2021 and customers were offered refunds up to £1,000 GBP to address the emotional and financial harm incurred. In 2024, it was found that only about 300 of 270,000 potentially affected customers submitted compensation claims (Hudspith 2024). Nova Scotia initially offered two years of free credit monitoring, which has now been extended to five years for all current and former customers (Nova Scotia 2025). Interestingly, there were limited responses made regarding the data leak that happened in Texas, with no formal customer notification, compensation or independent review.

Confronted with such "accidents," we ask who is more susceptible to having their energy data comprised. It is not uncommon for people to obtain a modest car insurance discount of approximately $20–30 per month in exchange for providing the company with detailed personal information, including driving habits and GPS location. Although this trade-off may appear worthwhile, individuals with more disposable wealth are more likely to refuse this exchange, deeming their personal data to carry a value far greater than such a discount. This example illustrates how individuals weigh the worth of their personal information differently depending on their economic circumstances, raising the question of whether similar inequalities shape decisions about energy data sharing and the attendant risks of breach.

There are few studies that focus on this topic—energy data has not been a center of attention of the energy transition literature. It is not surprising that the topic of energy data justice has been somewhat neglected. Nevertheless, some studies attempt to show the inequalities that persist with smart technologies—although not specifically about energy data. Irwin and colleagues argued that residents of public or rental housing are more vulnerable to data breaches, as they typically have less agency in determining which meters are installed, how their data are collected, and what privacy safeguards are in place, in contrast to homeowners (Irwin et al. n.d.). Vigurs et al. (2021) claimed that low income and vulnerable people tend to feel more monitored in their everyday life compared to the general population and that smart technologies do not erase vulnerability and actually have a tendency to raise privacy or surveillance concerns (Shirani et al. 2020). Furthermore, it is not difficult to infer that individuals with fewer financial resources, limited time, or reduced digital literacy may also struggle to fully understand the implications of a breach or to navigate the processes required to seek redress. Moreover, the disclosure of billing records or outstanding energy debts information risks reinforcing stigma and may be exploited by landlords, lenders, or insurers. Taken together, it is evident that the risks associated with energy data are not evenly distributed by instead mirror and potentially exacerbate existing social inequalities.

Later in the book, we draw a vision that we imagine for future energy systems in detail. And further, we discuss who should be responsible for energy data (see Chap. 7).

6.4 RI and Problem Shifting: Sustainability Challenges of Energy Data Storage

One of the greatest challenges of energy data storage is sustainability. Energy data offers tangible benefits for energy efficiency and grid management, it can facilitate greater integration of RE sources, improve demand forecasting, and enhance both billing accuracy and consumer control over energy use (Correa et al. 2017; Dyson et al. 2014; Hopf et al. 2018; Shahrokni et al. 2014; Sodenkamp et al. 2017). These advantages have been extensively examined in the literature. Yet, alongside these positive dimensions, it is equally important to discuss the potential drawbacks associated with long term energy data retention. In this chapter, we also discuss the less visible, and often less discussed issues around data storage and the sustainability impacts of data centers.

As discussed above, the demand for storage for energy data will increase over time and space, and the demand for data center use will likely increase too. In other words, energy data analysis requires longitudinal and granular big data, which is collected repeatedly over a long period of time. Storing a vast amount of energy data requires data centers that use significant amount of energy, water, and land and can also lead to waste, biodiversity issues (Al Kez et al. 2022; Ferreira et al. 2018; Sovacool et al. 2022).

The Lawrence Berkeley National Laboratory (LBNL) announced that data centers consume approximately 4.4% of electricity in 2023, which is expected to increase to 6–12% of total U.S. electricity consumption by 2028. Furthermore, they claimed that electricity consumption by data centers has grown markedly over the past decade—rising from an estimated 58 TWh in 2014 to 176 in 2023, and a further surge is expected: 325–580 TWh by 2028 (Shehabi et al. 2024). The International Energy Agency reported recently that global data center electricity consumption could roughly double to 945 TWh by 2030, approaching just under 3% of world electricity consumption. This indicates that consumption is growing by 15% annually through 2030, which is over four times faster than total demand growth in other sectors (International Energy Agency 2024).

The fast increase of data centers also impact water, which is essential for cooling. A single 1-megawatt data center can consume up to 25.5 million liters of water per year just for cooling, which is approximately equal to the daily water use of about 300,000 people in some contexts (Kaur 2025). A 2024 report published by the LBNL also shows that U.S. data centers alone use approximately 17 billion gallons of water for cooling and approximately 211 billion gallons are used indirectly via thermoelectric generation to supply their electricity (Shehabi et al. 2024).

The consequences of energy data storage have spatial and temporal dimensions. The spatial dimension of energy data storage concerns the quality of life of communities located proximate to data centers. Although data centers serve a wide array of digital functions beyond the storage and processing of energy data, their cumulative environmental impact underscores the urgent need to scrutinize the sustainability implications of energy data infrastructures. The expansion of data centers—both in physical size and in number—is closely linked with rising volumes of data across multiple sectors, and energy data plays a nontrivial role in this growth. Unlike other forms of data that may be subject to periodic deletion or archiving, energy data is often retained indefinitely under the rationale that continuous historical records are necessary for optimizing grid reliability, forecasting demand, and managing distributed energy sources. This practice contributes to escalating storage demands, which in turn drives increased energy and water consumption and a heavier carbon footprint. This latter point ties into the temporal dimension of energy data problem shifting. As a result, the governance of energy data must incorporate sustainability considerations alongside privacy and security, recognizing that the digital infrastructures underpinning low-carbon energy systems may themselves generate substantial environmental burdens if left unchecked.

Available estimates of data centers' energy and water use must be treated with caution, as the underlying information is fragmentary and often unreliable. Comprehensive data on the size, location, and operational characteristics of existing data center facilities are not publicly available, and figures that circulate in policy and research circles are rough approximations at best. This opacity is reinforced by corporate secrecy. Companies rarely disclose detailed information about their data centers, citing concerns over competitiveness and security. As a result, despite the increasingly severe environmental burdens associated with data center growth, society lacks even a precise accounting of their ecological footprint.

Data centers have increasingly been recognized as significant contributors to global carbon emissions (Al Kez et al. 2022; Ewim et al. 2023). As the scope and granularity of smart meter deployment expands, both in terms of sampling frequency and geographic reach, the volume of energy consumption data generated is projected to grow substantially, thereby intensifying the need for large-scale, long-term data storage infrastructure.

Yet, the environmental burden of energy data does not end with storage. As demonstrated in earlier sections, this data is routinely processed, modeled, and subjected to advanced analytics to support forecasting, demand management, and grid optimization, and generates even more data in the process. These computational tasks, many of which rely on energy-intensive high-performance or cloud-based systems, require substantial processing power and contribute to the overall environmental footprint of digital energy systems. This layered chain, from generation to storage to computation, reveals that the sustainability impact of energy data extends beyond mere electricity use and encompasses wider issues, including water consumption, heat generation, hardware demand (and related waste), and associated emissions. Therefore, energy data, although offering clear operational benefits, simultaneously introduces new ecological pressures that must be taken seriously in discussions of responsible energy data governance.

6.5 Who Should Be Responsible for Energy Data?

Determining who should bear responsibility for energy data is fundamental to ensuring privacy, security, and equitable governance in an increasingly data-driven energy system. Energy data are generated, stored, and used across multiple points, from smart meters and distribution networks to cloud platforms and market applications. No single actor can realistically guarantee their protection. Instead, accountability must be distributed and clearly defined among governments, utilities and ESCOs, private companies and technology providers, civil society organizations, and international standard bodies, with end users and data producers granted enforceable rights over their own information. Establishing such a shared but explicit framework of responsibility is essential to safeguarding public trust and enabling the ethical use of energy data in the energy transition.

6.5.1 Governments

At the macro-level of rule-making and policy design are governments. National and subnational governments have been foundational actors in defining the rules around how energy data are collected, stored, and used. They determine the frequency of data sampling, set obligations on utilities regarding who may access the data and under what terms, and mandate how data must be stored and protected. For example, Lee and Hess (2021) offer a comparative analysis of privacy policies in Europe and North America and show how different governments regulate smart meters in dimensions including customer opt-out or opt-in rights, data sampling intervals, sharing guidelines, storage responsibilities, and enforcement mechanisms such as independent oversight or regulatory agencies.

At the supranational level, the European Union (E.U.) has also been instrumental. The E.U.'s Mandate M/441, issued in 2009, set out requirements for creating open architectures and interoperability standards for utility meters (covering electricity, water, gas, heat) across member states (European Commission 2014). Under this mandate, European Standardization Organizations (CEN, CENELEC, ETSI) developed functional reference architectures and communication standards (for example, through TR 50572:2011) and produced harmonized requirements for privacy, security, and data interoperability (European Committee for Electrotechnical Standardization 2025).

More recently, governance in the E.U. has also expanded through the Smart Grids Task Force and the merged CEN-CENELEC-ETSI Coordination Group on Smart Grids (CG-SG), which advise on standards for interoperability, data access, and cybersecurity across smart metering and broader grid systems. These bodies help shape regulations such as those derived from the E.U. Clean Energy Package including implementing regulations on access to metering and consumption data. Together, national and supranational regulatory regimes thus control not only how meters are installed and managed but also how privacy is preserved: through laws (e.g., GDPR in the E.U.), technical standards (communication protocols, minimum security requirements), legal bases for data processing (consent, public interest), complaint mechanisms, and enforcement. These layers of regulation suggest that responsible governance of energy data is not only a technical matter but deeply political and institutional as well.

Within the E.U., country level governments are also pursuing their own data protection and security standards. For example, in Germany, the Energy Industry Act (EnWG) and the Metering Point Operation Act (MsbG) include rules about who may access data, under what conditions, and require that consumption data are stored only with necessary granularity and only for authorized purposes. More specifically, consumers are granted access to their own meter data, and there is a mandatory notification of data breach (Federal Ministry for Economic Affairs and Energy 2025).

Outside of the E.U., the U.K. have introduced the Smart Metering Data Access and Privacy Framework (DAPF) in 2018, which allows monthly data for billing processes, but daily data for any purpose is an opt out option, and half hourly data for any purpose is opt in only, which gives more options to consumers. Furthermore, distribution network operators can access smart meter data for regulated purposes only with a privacy plan and third parties can access smart meter data if there is opt in consent (Department for Business, Energy and Industrial Strategy 2018; Judson et al. 2025).

Despite these efforts, some countries lag behind in establishing privacy laws specific to energy data. In the U.S., the Department of Energy issued a Voluntary Code of Conduct that sets out principles of consent, notification, minimization, and security that utilities and third parties are encouraged to follow but are not mandatory, and South Korea is yet to have any smart meter specific regulations and their smart meter data falls under the general privacy law, Personal information Protection Act (PIPA).

Sustainability is another dimension that require accountability and liability related to energy data. Oh et al. (2025) provide a detailed review of emerging governmental regulations aimed at improving the sustainability of data centers, including measures to reduce energy consumption, water use, and electronic waste. Their analysis shows that while many jurisdictions have introduced policies to curb the environmental footprint of these facilities—such as efficiency standards, voluntary reporting schemes, and incentives for heat recovery—significant gaps remain. In particular, there is a continuing need to enhance the transparency of both energy and water use, so that operators, regulators, and the public can reliably assess actual performance. The absence of internationally harmonized goals or benchmarks further limits the comparability of sustainability efforts across borders and hinders collective progress. Moreover, the integration of renewable energy sources into data center operations is still uneven, with many facilities relying heavily on conventional electricity grids. Even where targets or standards exist, independent monitoring and enforcement are often weak or inconsistent, reducing their effectiveness. Taken together, these shortcomings underscore the importance of stronger, more coordinated governance if data centers are to support a genuinely sustainable digital infrastructure for the energy sector and beyond.

A central weakness of most government regulations concerning privacy or data centers is their lack of meaningful penalties and enforcement. Though consumers may have formal rights to lodge complaints, empirical studies show that regulators often fail to follow up, and violations frequently go unpunished. For instance, comparative reviews of global data privacy laws reveal that enforcement mechanisms are uneven and weak, even when statutes on paper look robust (Fuller 2017). In the U.S., watchdog agencies have repeatedly identified lax enforcement of privacy rules, with few cases resulting in real sanctions or operational changes by firms (Borner 2019). Additionally, ethical critiques of emerging infrastructure warn that governments sometimes knowingly overlook privacy breaches or environmental sustainability deficits under the banner of promoting innovation or energy transition (Dhirani et al. 2023). This regulatory reluctance creates a de facto tolerance for data injustice and allows powerful actors to externalize risk onto consumers and communities with minimal accountability.

To build a more credible and robust framework, governments need to move beyond generic data-protection or environmental provisions and develop dedicated legal instruments tailored to the distinctive risks of energy data. Such measures should define clear standards for data collection, storage, sharing, and deletion; mandate transparent reporting of the environmental footprint of energy-data infrastructures; and include enforceable penalties for non-compliance. By embedding these sector-specific requirements into law and providing mechanisms for regular oversight and public accountability, governments can strengthen both the privacy protections and the sustainability outcomes of the rapidly expanding energy data economy.

6.5.2 Utilities and ESCOs

The meso level involves voluntary action by companies and households. Therefore, the important player at this level are Utilities and ESCOs. Utilities and ESCOs, as the principal users and beneficiaries of energy data, bear a substantial share of responsibility and liability for its collection, storage, and management. Their operational, commercial and analytical reliance on detailed consumption information makes them central actors in the energy ecosystem. Indeed, it is precisely their capacity to aggregate, analyze and monetize this data that has brought issues of privacy, accountability, and governance to the forefront of contemporary energy debates. As custodians of highly sensitive information about consumers' energy use, household behavior, and even occupancy patterns, utilities and ESCOs must uphold rigorous standards of data protection, transparency, and ethical use. This entails not only compliance with regulatory requirements but also the proactive implementation of international governance mechanisms such as data minimization, purpose limitation, and independent auditing, to ensure that data are handled responsibly across their entire life cycle.

The importance of ESCO in maintaining and upholding privacy and cybersecurity of energy data is clearly evidenced by the requirements and frameworks that New York has introduced. In 2019, the New York Public Service Commission adopted new rules obligating utilities and ESCOs to uphold minimum security and data privacy protections for customer data—and if they were to fail, regulation allows utilities to deny or revoke their access to utility IT systems (Walton 2019). Similarly, New York also introduced a Data Access Framework, in which utilities are required to maintain data quality, ensure privacy protections, and company with standardized data access conditions (Wolfe 2023).

The Snohomish County Public Utility District (SNOPUD) in Washington State offers a valuable example of how utilities can integrate principles of responsibility and accountability into digital transformation. As one of the largest publicly owned utilities in the U.S., SNOPUD initiated the Connect Up program to modernize its grid infrastructure through the deployment of advanced metering systems and digital communication technologies. Recognizing that such innovations also heighten risks related to cybersecurity and data privacy, the utility adopted a precautionary approach by embedding protective measures into the program's initial design rather than treating them as afterthoughts. These measures included comprehensive vulnerability assessments, network segmentation, and continuous monitoring of data management systems. In doing so, SNOPUD demonstrated that technological advancement and data stewardship need not be in tension, but can be developed concurrently as complementary pillars of responsible energy innovation.

Another noteworthy example of responsible cybersecurity practice among utilities is provided by a major European distribution system operator (DSO) that collaborated with Industrial Defender to strengthen the resilience of its digital infrastructure. Serving millions of customers across a wide service territory, the DSO integrated comprehensive cybersecurity measures into its grid modernization strategy, emphasizing prevention, visibility, and accountability. Through the deployment of automated asset inventories, configuration monitoring, and Information Technology-Operational Technology (IT–OT) integration, the utility enhanced its capacity to detect and respond to anomalies in real time while ensuring compliance with both the Center for Internet Security Critical Security Controls and the E.U.'s Network and Information Security (NIS) Directive. ESCOs working alongside the utility were also required to adopt compatible security and privacy standards, creating a cohesive data protection environment across the energy ecosystem. This case exemplifies how utilities and their partners can assume active responsibility for safeguarding energy data by embedding cybersecurity as a foundational element of digital transition rather than treating it as a regulatory obligation.

In this sense, utilities and ESCOs have a pivotal position in recognizing and advancing the importance of privacy and cybersecurity within the energy data landscape. As the primary custodians and users of consumer energy information, they are uniquely situated to set industry standards and cultivate public trust in the digital energy transition. Yet, despite this strategic position, many utilities remain reluctant to engage fully with the ethical and technical dimensions of data governance, often perceiving cybersecurity and privacy compliance as external impositions rather than integral components of responsible energy management. This hesitancy not only exposes consumers to potential risks but also undermines the credibility and social legitimacy of innovation in the sector. To ensure a genuinely responsible transformation of energy systems, utilities and ESCOs must move beyond minimal compliance and adopt proactive, transparent, and participatory approaches to data stewardship—approaches that balance operational efficiency with respect for consumer rights and long-term societal trust.

6.5.3 Private Companies and Technology Providers

Private companies, including smart meter manufacturers, cloud service providers, and technology vendors, play an increasingly decisive role in ensuring the cybersecurity and privacy of energy data. As much of the digital infrastructure that underpins modern energy systems is now designed, maintained, or hosted by private actors, their practices directly shape the resilience and trustworthiness of energy data ecosystems. In recent years, several firms have begun to integrate privacy-by-design principles, embedding encryption, authentication, and access control mechanisms at the hardware and firmware levels of smart meters and extending these protections through secure cloud architectures (Gurses et al. 2011; Schaar 2010). These efforts reflect a gradual shift from reactive, compliance-based approaches to proactive, systemic integration of cybersecurity across the energy data value chain. By setting new technical standards and operational norms, private technology providers are not merely service contractors but key stewards of digital trust in the evolving landscape of energy governance.

A strong example of cybersecurity integration at the hardware and system architecture level can be seen in the collaboration between Thales and a major European utility during a large-scale smart meter rollout. In this project, Thales implemented a security infrastructure based on Hardware Security Modules (HSMs) and Public Key Infrastructure (PKI) to protect cryptographic keys and manage device authentication for millions of smart meters. The HSMs functioned as secure vaults for key storage and cryptographic operations, ensuring that private keys could not be extracted or manipulated even in the event of a system breach. Through PKI, each meter was issued a unique digital certificate, enabling mutual authentication between devices and utility servers and supporting the verification of firmware integrity. By enforcing compliance with high-assurance security standards such as FIPS 140-2 Level 3, the project established a robust "chain of trust" across the entire metering network. This case exemplifies how meter manufacturers and utilities can jointly institutionalize cybersecurity through technical and procedural safeguards, embedding privacy and data protection directly into the physical and digital infrastructure of energy systems rather than relying solely on post-hoc regulatory measures (Thales 2024, 2025).

A second example of responsible cybersecurity practice is provided by the collaboration between Cloudseed and a public utility undergoing a major digital infrastructure migration. The utility, seeking to modernize its operations and consolidate applications within a managed data environment, partnered with Cloudseed to redesign its cybersecurity architecture from the ground up. The project introduced next-generation firewalls, endpoint protection systems, intrusion detection, and a unified threat monitoring dashboard that enabled real-time visibility across operational systems. By integrating these capabilities, the utility reduced its average threat detection time from days to hours, strengthening both resilience and regulatory compliance. Moreover, the migration to a dedicated, securely managed cloud environment ensured better data isolation and more consistent enforcement of access controls. Importantly, the collaboration redefined the relationship between cloud providers and utilities—from one of passive hosting to one of shared responsibility for energy data protection. This case demonstrates how cloud service providers, when acting as active partners in security management, can contribute to a more accountable and transparent data ecosystem within the energy sector (Cloudseed 2025).

6.6 Reflections and Conclusions

Energy data straddle the liminal space where energy transitions and the information economy overlap. As such, they are highly sensitive to even minor disruptions or fluctuations in data collection, sharing, or storage. Data center owners and their power providers face several interrelated technical and economic challenges. They must guarantee high-availability (24/7) power with an extremely low probability of failure, ensure near-perfect power quality, and manage risk while minimizing both capital expenditures and operating costs. At the same time, they create substantial negative externalities in the shape of emissions and water pollution.

The rapid growth of digital services and cloud computing amplifies concerns about environmental impact, particularly greenhouse gas emissions (Ben-Shahar 2019) Innovations in cooling, architectural design, renewable energy sourcing, and hardware efficiency offer promising reductions in energy use (Ewim et al. 2023). A holistic approach—integrating technical upgrades with supportive policy measures and new regulatory perspectives on data-related externalities—is essential to reconcile data center needs with societal and environmental objectives.

To align data-center evolution with broader societal values, embracing anticipatory governance can help manage emerging risks and opportunities before they materialize. This involves foresight, broad stakeholder engagement, and iterative learning to shape policies, standards, and governance structures that anticipate data-related externalities, ethical considerations, and environmental impacts. By integrating foresight with practical governance mechanisms, organizations can steer innovation toward sustainable outcomes—balancing reliability and cost with public trust, privacy, and social welfare.

To that end, policymakers can implement Seventh Generation Principles by institutionalizing long-term sustainability and intergenerational equity as core values in decision-making processes (Da Costa et al. 2021). The Seventh Generation Principle is rooted in Haudenosaunee (Iroquois) philosophy and calls for decisions made today to ensure the well-being of the seventh generation into the future, highlighting responsibility for both the environment and social systems (Ebel and Rinke 2014). This principle challenges leaders and communities to prioritize long-term sustainability and justice, considering not only immediate needs but also collective impacts over time. Anticipatory governance complements this approach by encouraging proactive, forward-thinking policies that imagine possible futures and guide current action accordingly (Lysaght 2022; Umbach 2024). This can include establishing accountability mechanisms such as intergenerational impact assessments and ensuring the inclusion of youth and future generations in governance forums, for example by allocating seats to young representatives. Together, these frameworks promote equity, resilience, and the integration of diverse perspectives to shape sustainable outcomes.

References

Al Kez D, Foley AM, Laverty D, Del Rio DF, Sovacool B (2022) Exploring the sustainability challenges facing digitalization and internet data centers. J Clean Prod 371:133633. https://doi.org/10.1016/j.jclepro.2022.133633

Ben-Shahar O (2019) Data pollution. J Leg Anal 11:104–159. https://doi.org/10.1093/jla/laz005

BituoTechnik (2025) How long can energy data be stored? https://www.bituo-technik.com/kb/en-gb/how-long-can-energy-data-be-stored/

Borner P (2019) Government watchdog finds weak enforcement of US privacy regulations. https://thedataprivacygroup.com/us/blog/2019-3-8-government-watchdog-finds-weak-enforcement-of-us-privacy-regulations/

Cegal (2025) Elhub. https://www.cegal.com/en/dictionary/elhub

Cloudseed (2025) A utility company strengthens its cyber security measures. https://www.cloudseed.io/case-studies/a-utility-company-strengthens-its-cyber-security-measures

Correa E, Inga E, Inga J, Hincapié R (2017) Electrical consumption pattern base on meter data management system using big data techniques. In: 2017 International conference on information systems and computer science (INCISCOS), pp 334–339

Da Costa NG, Farias G, Wasieleski D, Annett A (2021) Seven principles for seven generations: moral boundaries for transformational change. Hum Manag J 6(3):313–328. https://doi.org/10.1007/s41463-021-00116-y

Department for Business, Energy and Industrial Strategy (2018) Smart metering implementation programme: review of the data access and privacy framework. Department for Business, Energy and Industrial Strategy, United Kingdom

Department for Business, Energy and Industrial Strategy (2021) Energy-related products policy framework

Dhirani LL, Mukhtiar N, Chowdhry BS, Newe T (2023) Ethical dilemmas and privacy issues in emerging technologies: a review. Sensors (Basel, Switzerland) 23(3):1151. https://doi.org/10.3390/s23031151

Dyson MEH, Borgeson SD, Tabone MD, Callaway DS (2014) Using smart meter data to estimate demand response potential, with application to solar energy integration. Energy Policy 73:607–619. https://doi.org/10.1016/j.enpol.2014.05.053

Ebel A, Rinke T (2014) Listening to the voices of young and future generations. In: State of the world 2014: governing for sustainability. Island Press/Center for Resource Economics, Washington, DC, pp 82–90

Elhub (2025) Elhub.No—Elhub. https://elhub.no/

European Commission (2014) Mandate M441 for smart meters, March 2009. European Commission. https://energy.ec.europa.eu/publications/mandate-m441-smart-meters-march-2009_en

European Committee for Electrotechnical Standardization (2025). Smart meters. https://www.cencenelec.eu/areas-of-work/cen-cenelec-topics/smart-grids-and-meters/smart-meters

Ewim DRE, Ninduwezuor-Ehiobu N, Orikpete OF, Egbokhaebho BA, Fawole AA, Onunka C (2023) Impact of data centers on climate change: a review of energy efficient strategies. J Eng Exact Sci 9(6):16397-01e. https://doi.org/10.18540/jcecvl9iss6pp16397-01e

Federal Ministry for Economic Affairs and Energy (2025) Smart metering—subject to stringent data protection and security rules. Federal Ministry for Economic Affairs and Energy

Federal Office for Information Security (2025) Smart metering—subject to stringent data protection and security rules. Federal Ministry for Economic Affairs and Energy, Germany

Ferreira J, Callou G, Josua A, Maciel P (2018) Estimating the environmental impact of data centers. In: 2018 IEEE 17th international symposium on network computing and applications (NCA), pp 1–4

Fuller CS (2017) The perils of privacy regulation. Rev Austrian Econ 30(2):193–214. https://doi.org/10.1007/s11138-016-0345-0

Greig J (2025) Nearly 300,000 people were impacted by cyberattack on Nova Scotia Power. https://therecord.media/thousands-impacted-cyber-nova-scotia

Gurses S, Troncoso C, Diaz C (2011) Engineering privacy by design. Priv Data Prot

Hopf K, Sodenkamp M, Staake T (2018) Enhancing energy efficiency in the residential sector with smart meter data analytics. Electron Mark 28(4):453–473. https://doi.org/10.1007/s12525-018-0290-9

Hudspith J (2024) Hundreds of thousands yet to claim refund from bust energy firm. https://dbsne.com/post/hundreds-of-thousands-yet-to-claim-refund-from-bust-energy-firm

International Energy Agency (2024) Energy demand from AI

Irwin G, Hurst A, Rollins S (n.d.) Who has my data? Illuminating renters' smart meter privacy concerns. https://cups.cs.cmu.edu/soups/2014/posters/soups2014_posters-paper13.pdf

Judson E, Reynolds F, Chitty M, Johnson T, Jenkins D (2025) Data protection and smart meter data—open energy. https://openenergy.org.uk/data-protection-and-smart-meter-data/

Kaur D (2025) Cloud's hidden cost: data centre water consumption creates a global crisis. https://www.cloudcomputing-news.net/news/data-centre-water-consumption-crisis/

Kleinman Z (2020) People's energy data breach affects all 270,000 customers. BBC, December 17

Lee D, Hess DJ (2021) Data privacy and residential smart meters: comparative analysis and harmonization potential. Util Policy 70:101188

Lysaght T (2022) Anticipatory governance and foresight in regulating for uncertainty. Am J Bioeth 22(1):51–53. https://doi.org/10.1080/15265161.2021.2001111

Matheson D, Jing C, Monforte F (2004) Meter data management for the electricity market. In: 2004 International conference on probabilistic methods applied to power systems, pp 118–122

Nova Scotia (2025) Cyber. https://nspower.ca/home---cyber

Oh N, Son E, Lee D (2025) Data centers' environmental challenges: a comparative review analysis and policy harmonization

Perrow C (2011) Normal accidents: living with high risk technologies, Updated edn. Princeton University Press, New Jersey

Schaar P (2010) Privacy by design. Identity Inf Soc 3(2):267–274. https://doi.org/10.1007/s12394-010-0055-x

Shahrokni H, Levihn F, Brandt N (2014) Big meter data analysis of the energy efficiency potential in Stockholm's building stock. Energy Build 78:153–164. https://doi.org/10.1016/j.enbuild.2014.04.017

Shehabi A, Newkirk A, Smith SJ, Hubbard A, Lei N, Abu Bakar Siddik Md, Holecek B, Koomey J, Masanet E, Sartor D (2024) 2024 United States data center energy usage report. Lawrence Berkeley National Laboratory. https://doi.org/10.71468/P1WC7Q

Shirani F, Groves C, Henwood K, Pidgeon N, Roberts E (2020) 'I'm the Smart Meter': perceptions of smart technology amongst vulnerable consumers. Energy Policy 144:111637. https://doi.org/10.1016/j.enpol.2020.111637

Sodenkamp M, Kozlovskiy I, Hopf K, Staake T (2017) Smart meter data analytics for enhanced energy efficiency in the residential sector. In: Wirtschaftsinformatik 2017 proceedings. https://aisel.aisnet.org/wi2017/track12/paper/10

Sovacool BK, Monyei CG, Upham P (2022) Making the internet globally sustainable: technical and policy options for improved energy management, governance and community acceptance of Nordic datacenters. Renew Sustain Energy Rev 154:111793. https://doi.org/10.1016/j.rser.2021.111793

SSE Energy Solutions (2025) Smart data guide. SSE Energy Solutions. https://www.sseenergysolutions.co.uk/smart-data-guide

Thalen M (2021) Exclusive: hacker reveals smart meters are spilling secrets about the Texas snowstorm. https://www.dailydot.com/debug/hacker-smart-meter-texas-snowstorm/

Thales (2024) Enabling secure connectivity for smart meters

Thales (2025) Smart meter security for European energy utility—case study. https://cpl.thalesgroup.com/resources/encryption/smart-meter-security-european-energy-utility-case-study

UK Power Networks (2015) Smart meter energy consumption data in London households. London Datastore. https://data.london.gov.uk/dataset/smartmeter-energy-use-data-in-london-households/

Umbach G (2024) Futures in EU governance: anticipatory governance, strategic foresight and EU better regulation. Eur Law J 30(3):409–421. https://doi.org/10.1111/eulj.12519

United Kingdom Office for National Statistics (2025) Families and households in the UK. Office for National Statistics. https://www.ons.gov.uk/peoplepopulationandcommunity/birthsdeathsandmarriages/families/bulletins/familiesandhouseholds/2023

US Census Bureau (2024) Historical households tables. https://www.census.gov/data/tables/time-series/demo/families/households.html

Vigurs C, Maidment C, Fell MJ, Shipworth D (2021) Privacy and data sharing in smart local energy systems: technical report. EnergyREV

Walton R (2019) New York adopts utility-ESCO cybersecurity standards, rejects insurance requirements. Utility Dive. https://www.utilitydive.com/news/new-york-adopts-utility-esco-cybersecurity-requirements-rejects-insurance/565333/

Wolfe S (2023) Case study: how a utility built security into its digital transformation. https://www.renewableenergyworld.com/power-grid/grid-modernization/case-study-how-a-utility-built-security-into-its-digital-transformation/

Part III
How Far Can We Take Energy Data Insights?

Chapter 7
Sociotechnical Imaginaries of Energy Data

7.1 Our Energy Data Future

Can data be *owned*? Rachur et al. (2022) argue that data indeed is personal property because (1) data has a precise definition provided by the General Data Protection Regulation (GDPR), (2) data can be exclusively controlled by owner and can be given, shared, or sold, and (3) data has legitimate claim to exclusivity. In other words, in legal terms, data can fall under the category of personal property—yet its treatment in everyday practice tells a very different story. In short, data is rarely regarded as something over which individuals exercise clear ownership or control, despite strong arguments that it should be (Al-Khouri 2012).

With the rapid expansion of digital infrastructures and the proliferation of the Internet of Things, data flow has become deeply embedded in daily life, often occurring without conscious recognition. Most individuals remain unaware of the circumstances under which their personal data is extracted, the purposes for which it is used, the duration (and location) for which it is stored, and the forms of analysis to which it is subjected. This disconnect between the legal definition of data and its practical handling exposes asymmetry of power between data subjects and data collectors and users, and highlights the urgent need for more robust frameworks that align property rights vis-à-vis personal data. This disconnect is especially pronounced in the case of energy data. Although meter data is indispensable for the operation and modernization of energy systems, public access to it is highly restricted often limited to partial and carefully curated datasets, if it is available at all. Most consumers know little about the frequency with which their own energy data is collected, and the procedures—if they exist—for requesting deletion. The institutions that hold and manage such data are typically utilities, regulatory bodies, or researchers with special authorization (Webborn et al. 2019), while the individuals who generate the data through their everyday energy use frequency face barriers to accessing it themselves.

Even in contexts where companies provide consumers with portals or applications to view their household consumption, such access is constrained to visualization

L. A. Nguyen Long and D. Lee, *Energy Data*, Lecture Notes in Energy 46,
https://doi.org/10.1007/978-3-032-19171-7_7

rather than control. Consumers are generally unable to rectify errors, request deletion, or influence decision about how their data is stored, analyzed, or shared. As Hummel et al. (2021) put it, the socio-cultural and the political recognition of data subjects is missing from the energy data frameworks.

This asymmetry underscores a broader social tension: although energy data is framed as essential for advancing efficiency, sustainability (to an extent), and innovation, the rights and agency of those who produce it remain narrowly defined. In effect, consumers are positioned not as owners of their own energy data but as passive sources, reinforcing the imbalance between data subjects and the institutions that profit from or regulate digital energy infrastructures. To address this imbalance, we advance an alternative approach grounded in the notion of data stewardship—a framework that emphasizes responsibility, care, and accountability over mere possession. Through data stewardship, energy data can be managed in ways that respect individual rights while recognizing the collective value of data for system optimization, sustainability, and public benefit.

7.2 Sociotechnical Imaginaries and Collective Foresight

We use the sociotechnical imaginaries, a concept developed by Jasanoff and Kim (2009, 2013), to provide a conceptual foundation for the collective foresight and a multidimensional governance approach. Sociotechnical imaginaries refer to collectively held visions of how technology and society should co-evolve in the future. In principle, these imaginaries are coherent enough to inspire shared commitments and translate into concrete policies and institutional arrangements. Nevertheless, in practice, they are often marked by contestation and fluidity, leaving considerable space for interpretation, negotiation, and political struggle over what futures should be pursued and by whom (Jasanoff and Kim 2009, 2013; Schiølin 2020; Sismondo 2020). Within these imaginaries, notions about the appropriate societal functions of technologies—like the assemblages that constitute renewable energy—and the role of the state vis-à-vis these technologies become deeply intertwined.

The concept sociotechnical imaginaries is useful because that is precisely what we are doing—imagining. We are imagining a future, where energy data is not commercialized or commodified. Instead, we hope for energy data to be managed as a public good, insulated from extractive profit logic. We aim for a future, where energy data are governed by responsible individuals at the micro-level, embedded within ethically grounded and context-sensitive systems at the meso-level, and aligned with a cohesive and globally informed governance framework at the macro-level. A second reason for which imaginaries are crucial for examining energy data is because of their explicit emphasis on the importance of values in shaping how we understand data and its societal role.

Sociotechnical imaginaries are especially relevant to energy data because its governance, as explored in earlier chapters, implicates far more than questions of technical performance or market efficiency. Energy data is inseparable from concerns about individual rights, distributive justice, and environmental accountability, and thus demands a framework that foregrounds these levels rather than relegating them to secondary considerations. A sociotechnical imaginary provides such a framework by enabling the articulation of a shared vision that draws together diverse actors, systems, and institutions. Ergo, it supports a multilevel approach that places democratic values and public interest at the center of energy data governance, rather than reinforcing existing inequities or vulnerabilities.

7.3 Sociotechnical Imaginaries of Responsible Energy Data

Energy data governance is situated contemporaneously in three dimensions: micro-, meso-, and macro, which contribute to the sociotechnical imaginary of energy data. At the micro level, the practice of data donation represents an emerging avenue through which individuals can exercise agency over their energy data, contributing it voluntarily to research or public-interest initiatives rather than having it passively extracted. This concept is gaining traction in the digital privacy literature as a potential alternative to conventional, often opaque forms of data collection, as it centers consent, transparency, and user autonomy.

At the meso level, we situate energy data within the the development of consumption corridors. These corridors acknowledge that some minimal level of energy data must be consumed, and that while there are individuals and firms that over-consume energy data, there are constituencies who do not produce sufficient energy data to serve their energy transition needs (e.g., the energy poor). This challenges us to consider what types of data are collected, with what frequency, as well as for what purpose and how are they collected. Through the lens of consumption corridors, data stewardship thus becomes an ethical relationship, one that obliges institutions and intermediaries to act with care toward the individuals and communities whose lives the data reflect.

At the macro level, the concept of Imaginary Future Generations (IFGs) signals a move toward transnational governance structures capable of institutionalizing these principles. Such councils would enable the articulation and coordination of a global agenda for ethical energy data practices, embedding norms of responsible data collection, storage, and use into regulatory frameworks and international cooperation. IFGs further calls for the continuous updating of energy data governance to allow for the need to accompany rapid developments in the field of energy. Through these interlocking levels, we envision a sociotechnical imaginary in which energy data governance is collectively shaped by individual agency, institutional ethics, and global foresight.

As illustrated in Fig. 7.1, these three levels are mutually reinforcing rather than discrete or sequential. Although they may seem hierarchical with micro at the bottom and macro at the top, they are coupled by the presence of energy data across all dimensions rendering the relationships between them mutually reinforcing and interdependent, with each dimension shaping and being shaped by the others in a dynamic process of co-construction. At the micro level, the practice and ethos of data donation foster both individual awareness and a sense of agency, laying the groundwork for setting the boundaries for consumption corridors. In turn, setting thresholds for data sufficiency not only validates data donation as a legitimate and ethical mode of engagement but also embeds it within shared social norms and collective expectations. This meso-level culture provides the normative and institutional context necessary for the macro-level work of the Imaginary Future Generations (IFGs), which seek to formalize these values into durable governance frameworks. Conversely, the IFGs help consolidate and scale the ethical commitments developed at the micro and meso levels, transforming them into policy instruments and transnational standards. Finally, the continued participation of individuals through data donation offers a crucial source of legitimacy and accountability for governance efforts, ensuring that institutional action remains grounded in public consent and democratic support. This model offers a pathway for reimagining energy data as a socially embedded and ethically governed resource and for developing viable alternative sociotechnical imaginaries around energy data. Below, we elaborate on each dimension in greater depth, examining the specific mechanisms, practices, and institutional arrangements through which a more just and accountable energy data future may be realized.

7.4 Imaginaries in the Micro: Data Donation and Gift Relationship

Energy consumption data collection is a paradoxical issue. On the one hand, it addresses certain aspects of social and environmental challenges: a greater control over the energy bill, which can be beneficial to lower income households, environmental benefits that come with more efficient and systematic grid management and energy technology management, and an easier transition towards RE. On the other hand, energy consumption data collection leads to significant problems such as privacy risks and contributing to climate change by storing masses amount of data. Data donation by energy users may help ease these tensions (Lee et al. 2024).

Data donation is a form of public participation in which "people voluntarily contribute their own personal data that was generated… to a collective dataset" (Bietz et al. 2019:1). The concept has been discussed in the health and medicine literature to encourage and publicize the importance of quality data for academic and medical research (Bietz et al. 2019; Krutzinna and Floridi 2019; Skatova et al. 2014; Strotbaum et al. 2019). Some studies present data donation as a form of ethical behavior that supports the greater good (Krutzinna and Floridi 2019). Others discuss posthumous medical data donation as a way to promote scientific research and to increase accessibility to patient and medical data (Krutzinna et al. 2019; Shaw et al. 2015).

As digital data become ubiquitous, the notion of data donation has been adapted to digital PII. Social justice and welfare are among the key foundations for digital

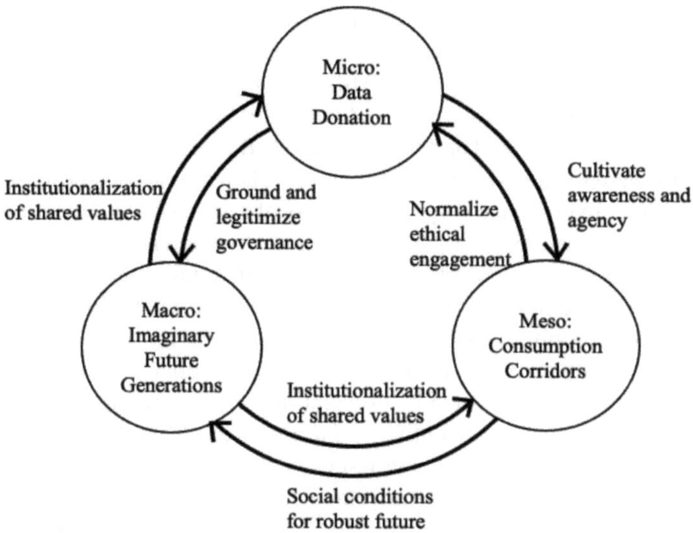

Fig. 7.1 Complementary of the three dimensions. *Caption: The diagram illustrates how three levels of sociotechnical imaginaries reinforce one another in shaping an environmentally and socially sustainable society. At the micro level, practices such as data donation depend on individual awareness and a willingness to engage ethically. These actions help support and legitimize broader forms of governance. At the macro level, the idea of imaginary future generations serves as a guiding perspective that encourages the creation and stabilization of shared values. These values flow into the meso level, where they inform the development of consumption corridors. In turn, these corridors shape everyday conditions and help nurture and normalize ethical engagement for the micro level and it provides social conditions for robust future for the macro level. The circular arrows connecting all three levels show how each sphere continually influences the others, creating an ongoing process through which shared values become established and a more resilient future can be built*

data donation. A few studies claim that data donation is a type of privacy preservation framework that can ensure high quality social science research (Boeschoten et al. 2022), as it provides direct access to relevant data. In the digital data donation literature, digital trace data and computational social science become particularly relevant for studying human behavior (Araujo et al. 2022; Boeschoten et al. 2022), and some studies claim the importance of donating real time data by using specific software or plugin that monitor user behavior (Christner et al. 2022). Here, we argue that energy data donation, which we embed in the framework of gift relationship, can be a solution to dual problems of privacy and emissions that energy data collection and storage face.

The *gift relationship* goes further than data donation because it highlights the absence of profit, capitalistic exchanges, and market. Initially introduced by Titmuss (1970)—and highly cited in philosophical and genetics research (Busby 2004; Locock and Boylan 2016; Rapport and Maggs 2002; Tutton 2002)—to accentuate the importance of blood and organ donation that is free from market intervention and the logic of capitalism, the concept of gift relationship stresses the importance of altruism and the public good. Titmuss claimed that blood and organ donation should be a gift that does not expect a financial return as one would in a contract relationship or in an exchange of commodity. Titmuss's main concern was with equity of accessibility of blood and organs and the financially motivated third parties creating and intervening with the blood and organs market (Rapport and Maggs 2002). These ideas carry forward to other domains, like energy. The gift relationship aligns with the inclusionary, anticipatory, and responsive tenets of responsible innovation. By positioning the energy consumer as the gift giver, data donation is inclusive and requires responsiveness to donors' preference. Moreover, it requires energy consumers to anticipate what can be done with the data before they decide to donate it.

The gift relationship is useful in imagining a new paradigm in the field of energy studies, which currently is highly motivated by financial incentives through various systems and policies such as prosumerism, net metering, financial savings of customers, and monetary incentives (Gautier et al. 2019; Lee and Hess 2021; Lee et al. 2020). It also addresses the issues of data justice by addressing notions of community creation, voluntary movements, and civic participation, which are central to building gift relationships (Boulet et al. 2008). By engaging donors in the gift relationship, data donation touches on the recognitional, procedural, and distributional dimensions of justice acknowledging donors' roles in data production and consumption, intensifying their participation in this process, and respecting their self-determination around issues of own-data.

We couple the concept of gift relationship with energy consumption data donation with two goals in mind: (1) to provide potential solutions to privacy and environmental concerns that are associated with the collection and storage of energy consumption data, (2) to put forward alternative imaginaries of energy data. Reflecting the essences of the gift relationship, data donation contemporaneously addresses the underlying intention of altruism and the desire to serve the public (Lee et al. 2024). Essentially, we argue that for data donation to be widely disseminated and promulgated, the concept must embed the ideas of compassion, selflessness, and social and public kindness.

7.4.1 The Right to Rectification and Deletion

In this section, we turn to the question of the right to rectification and delete energy data, which is fundamental to establishing fair and accountable energy data governance. A one-time act of consent to share or "donate" personal data should not be treated as an indefinite agreement. Rather, individuals must be kept abreast of how their data is being stored, analyzed, and potentially shared over time. Consent must be understood as an ongoing and revisable process, not a permanent transfer of rights.

The mere act of providing energy data to a utility company or energy service provider does not extinguish the rights of the individual who generated that data. Upholding rights of rectification and deletion is critical for reinforcing individual agency and establishing a more balance framework of data stewardship. Data subjects should retain the ability to withdraw from their donation if they prefer. The right to request deletion ensures that individuals are not locked into data donation agreements that may no longer reflect their preferences or serve their interests. This principle extends naturally to the question of ownership. Treating consent as a transfer of ownership risks consolidating power in the hands of utilities and ESCOs, while disempowering the very people whose data underpins energy systems.

Safeguarding the right to rectification and deletion recognizes that continued access to energy data is contingent upon ongoing trust. This will foster a relationship of accountability in which firms must remain attentive to the preferences of data providers, rather than treating consent as a one-time transaction. In doing so, utilities and energy service provides are reminded that data is not an entitlement but a conditional resource, granted through the willingness of individuals to share it.

This is where the donation of energy data is fundamentally different from the donation of blood and organs. Biological donations are finite and serve a direct, often immediate purpose, after which their use is concluded. By contrast, energy data can be stored, repurposed, and analyzed indefinitely, creating ongoing implications of privacy, sustainability, and ownership. Precisely because of this potential for extended use, the continuous monitoring of how energy data is managed is essential. Without such oversight, data donors risk losing meaningful control over information that remains valuable—and potentially vulnerable—long after the point of collection.

In the Netherlands, data governance practices surrounding smart meters provide an instructive case. Customers are given the option to actively opt in to more granular forms of energy data collection, which can be viewed as a form of voluntary data donation (Aben et al. 2021; Lee and Hess 2021). This framework reflects an acknowledgement of individual choice and consent in the management of household energy data. Yet, even within this comparatively progressive model, significant limitations remain. Although individuals may decide whether to participate there is little transparency regarding why energy data must be collected, how long it is stored, or what specific consequences may follow from agreeing to collection. Moreover, once consent is given, customers lack the right to rectify errors or request deletion of their energy data. In practice, this means that Dutch customers exercise agency at the point of entry, and have little control thereafter, which reinforces the power asymmetry between data donors and the institutions that manage and profit from their information.

7.4.2 *Privacy and Sustainability of Data Donation*

Data donation can resolve privacy issues, which are so closely entangled with energy data collection. Because data donation relies on the good will of the public and affords them control over data collection as the donor, organizations that handle energy consumption data are more likely to limit the amount of collected data to what is truly essential (i.e., the principle of data minimization), and to use data only for the specified purposes (i.e., purpose minimization and use limitation). In a similar fashion, increased accessibility improves processes of data rectification and erasure. Thus, data donation could better align energy data collection with the privacy principles proposed by the Fairness in Information Practices Principles (FIPPs) and the EU General Data Protection Regulations (GDPR). In this sense, data donation honors privacy regulations and guidelines proposed in both sets of principles.

Furthermore, obtaining energy consumption data via data donation could yield long-term benefits. The processes and ideas of data donation will ultimately change societal perception of data as something that can be owned and controlled by individuals. This returns the agency of data collection processes back to the public and places the energy organizations as secondary and passive recipients of data. We hope that this alternative imaginary of energy data collection stimulates the design and development of new safeguards consumers' privacy and that allows for alternative thresholds for individual and public ownership of data. This enables the development, establishment, and implementation of appropriate governance mechanism which emphasize transparency and prevent the misuse or mishandling of data. These mechanisms have the added benefit of raising donor trust and engagement in the process. Donors' active involvement in the deliberations and design of these mechanisms—for instance through co-creation (Van der Graaf et al. 2021)—will further efforts to ensure that they are democratically achieved.

Data donation also has the potential to mitigate the aforementioned energy data climate related challenges. Data donation is likely to generate more complete energy data that is higher in quality (i.e., lower rates of missingness, less noise), reducing the amount of data needed to make accurate predictions. This also reduces the size of data that must be stored. Consequently, energy models can achieve comparable if not higher levels of accuracy with less data. This in turn will lower demand for data storage and management.

7.5 Imaginaries in the Meso: Consumption Corridors

Much of the literature on energy data make broad claims about what data can be collected, and with what frequency. This capacity to collect "big energy data" is often described in celebratory language which emphasizes the energy efficiency and sustainability gains that these data help to deliver. This can be seen in both the academic literature (see Chaps. 4, 5, and 6) and in the print media (see Chap. 3). Rarely, if ever, are the risks of energy data considered side by side with their gains.

The existence of risks does not preclude the use of a technology. Most common technologies from household appliances to bullet trains contain some risk, and their use is often predicated on the assumption that (1) the risks have been sufficiently minimized and (2) the benefits of using the technology far outweigh the risks it entails. And when it comes to energy data, there is evidence that suggests that the planetary risk of failing to transition to renewable energy are considerable. Because energy data is critical to the adoption of renewable energy, some measure of energy must be collected. At the same time, we know that some households are able to opt out of energy data production, while others do not even have the chance to opt in. How then can we determine the right levels of energy data to collect in order to meet the energy transition needs in an ethical and just fashion? The concept of consumption corridors offers us a solution that addresses both energy data insufficiency and energy data superfluousness simultaneously.

The concept of consumption corridors emerged from sustainability scholarship to reconcile the tension arising from a paradox of contemporary society: overconsumption threatens to push us beyond our planetary boundaries into a climate emergency even as a substantial portion of earth's human inhabitants live well below the poverty line. For instance, a recent report released by Oxfam (2025:1) claims that "since 1990, the richest 0.1% have increased their share of total emissions by 32%, whilst the poorest half of humanity have actually seen their share fall by 3%" the report further claims that "To stay within the limits of the 1.5 °C threshold, the richest 0.1% would need to cut their per capita emissions by 99% by 2030." In light of this paradox, Fuchs and colleagues (Fuchs et al. 2021) develop a framework that reconciles the need to increase consumption among the earth's poor in order to allow them to achieve a dignified existence while pushing for regulations that substantially reduce the carbon emissions of high-emitting individuals and organizations. Their concept is called: consumption corridors.

Consumption corridors occupy that liminal space which is bounded on the bottom end by a minimum threshold, above which every individual can live a good life, and a maximum threshold which sets the limit for sufficient use of resources. Fuchs et al. (2021) further argue that it is nation states and the international community that have the responsibility of ensuring that all generations across time and space can meet these needs by the setting of minimum and maximum standards. Since individual needs and circumstances can vary greatly, it is important to critically engage with the philosophies and ethics through which minimum and maximum standards of resource use are determined. A first step would be to critically interrogate what is meant by "the notion of a good life."

The question of what is required in order to ensure that each individual can live a good life is wrapped up in notions of ideas of justice and freedom. For instance, as Di Giuglio and Fuchs point out, the capabilities approach argues that all humans possess intrinsic capabilities (Nussbaum 2013). These capabilities include the ability to form emotional attachments, having some measure of control over one's environment, the ability to use and regulate sensual functioning, and health and bodily integrity, so as not to die prematurely. Building on this basic premise, justice entails that each individual must have the opportunity to develop the needed capabilities to pursue a good life (Nussbaum 2013).

Another understanding of a good life shifts away from human capabilities to focus on "objective human needs" (Clair 2017; Costanza et al. 2007; Max-Neef et al. 1989).A good life is achievable, under this view if the existing "social, built, human and natural capital" provide each individual with the sufficient conditions to meet their objective needs, at a minimum, and their level of satisfaction with how those needs are met. Thus, objective needs provide the minimum threshold for a consumption corridor that is sustainable so long as it enables humans in both the present and the future to meet their objective needs (Fuchs et al. 2021). While this may—to some extent—take care of the lower threshold of the corridor, it does not provide guidance on how to set the upper threshold.

Sustainable development and justice are guiding principles for setting the maximum threshold. They require us to consider present and future distributions of scarce resources, and the quality of these resources. It is also important that the means through which resources are accessed and distributed limit the harm done to earth biomes. Following this logic, once energy data has achieved its functions (i.e., enabling capabilities and achieving objective needs related to efficiency, intermittency, flexibility and security), its continued existence is no longer justifiable and its continued storage has distributive consequences; therefore, it must be destroyed. Returning to the notion of sociotechnical imaginaries, the setting of consumption corridors is described by some as a prerequisite for creating an alternative world where, "people enjoy individual freedoms while prioritizing societal wellbeing; natural resources are managed so as to meet finite needs and not endless greed, for all living beings, now and in the future" (Sahakian et al. 2021:1).

7.5.1 Why We Should Set Minimum Standards for Energy Data

In previous chapters, we have explored how energy data collection has been justified by arguments about its contributions towards achieving an energy transition and towards monitoring and addressing energy poverty by detecting underconsumption of energy (Chaps. 1, 2, and 3). By some estimates, between 1 and 3 billion people do not have enough energy to meet their cooking and illumination needs (Guruswamy 2011). The implication of such a shortfall of consumption for their social and economic well-being is clear. Thus, it will be difficult for them to achieve a good life without energy data. A first step would be to ensure that these data are collected via methods that respect these household's data rights. One solution might be to set the minimum threshold for the consumption corridor at the amount of data individuals are willing to donate voluntarily via a gift relationship. A second condition may regard the specification of its uses, the objective needs those uses contribute to acheiving, and the capabilities specific data enable in order to avoid the frivolous generation and collection of energy data.

7.5.2 Why We Should Set Maximum Standards for Energy Data

The maximum level must be one above which the risk of present and future environmental and social harms outweigh present and future gains from the production, sharing and storage of energy data. Although energy data use continues to intensify, there are good reasons to reconsider the current systems of data center collection, storage, and distribution. The massive amounts of energy data needed to enable energy efficiency, flexibility, and decentralization come with sustainability risks related to data storage. According to the most recent estimates by the International Energy Agency, data centers account for approximately 1–1.5% of global energy use (International Energy Agency 2025). Despite substantial efforts to shift to increase these structures' energy efficiency and to switch to renewable energy sources, the report suggests that data centers continue to rely heavily on fossil fuels (Thangam et al. 2024). As the number of data centers continues to surge due to advancements artificial intelligence, this dependency relationship is expected to intensify. Data centers reliance and use of water resource is also an increasing concern (Bux et al. 2025; Siddik et al. 2021). Both trends related to energy data storage have implications for present and future generations, calling for an anticipatory governance approach. A focus on maximum thresholds does not contemplate the maximum amount of data to produce, it also challenges us to imagine and anticipate conditions under which energy data should be retired or abandoned.

Another reason to reconsider the systems that grow our reliance on energy data is privacy. Energy data comes with privacy risks due to methods of distributing data, some of which are done without consent and in order to financially capitalize on energy data does violating the data rights of households. This has recognitional justice implications, calling for greater reflexivity. Moreover, energy data comes with inequity risks, especially in systems where the most socioeconomically marginalized cannot opt out of energy data collection. Here, procedural and distributional justice are dually violated, calling for greater responsiveness and participation by those who are affected. However, setting the upper limit for energy data consumption can be a tricky and complicated process. This requires participatory, multi-stakeholder forums that uphold principles of responsible innovation.

7.5.3 The Need to Set Consumption Corridors by Following the Responsible Innovation Principles

The preceding discussions on the environmental and social risks of energy data puts into full view the limitations of technology centric approaches to sustainability challenges. This calls for a responsible innovation approach that can allay some of the risks and costs associated with energy data so that we can avoid a vicious circle of more iterative cycles of technologies being developed to solve the problems emerging from technologies developed to solve technology-derived problems of a previous generation. Solutions must then also consider social dynamics which shift the emphasis from one's side of the energy system (production) to another (consumption).

Most of the energy transition solutions focus on the production side of the energy system, emphasizing production processes that can be decentralized, made more reliable, efficient, and flexible. However, the recently emerging research on energy citizenship is slowly shifting the lens to solutions lying on the consumption side of the equation (Lennon et al. 2019; Long et al. 2024; Wieczorek et al. 2024). For energy data, the consumption corridors concept align with the idea that sustainability solutions may also be addressed from the demand side of the equation (Di Giulio et al. 2021; Murphy and Cohen 2001), in ways that reconcile energy justice considerations related to energy poverty and environmental degradation.

Following responsible innovation, the negative environmental and social justice implications of energy data call for an inclusive and participatory, anticipatory, and responsive approach to energy data governance. Although responsible innovation offers the methodology for deciding collectively on energy data consumption corridors, which offers an organizing principle around which such discussions can take place. More specifically, the focus on eudemonia, or "a good life," encourages us to think energy data consumption corridors as a pathway to an imagined future where we can all achieve a good life via principles of sufficiency, freedom, and justice where all humans can meet their basic needs because resources are not cordoned off in the hands of a few, powerful individuals (Jasanoff and Kim 2013).

7.6 Imaginaries in the Macro: Participation by Future Generations

At the macro level, we situate a concept that emphasizes the necessity of representing the interests of future generations within present-day governance structures. This perspective draws attention to long-term, systemic challenges (i.e., climate change, ecological degradation, and resource depletion) that extend far beyond the temporal horizons of conventional policy agendas, which tend to prioritize immediate economic growth or short-term social equity. By integrating future-oriented thinking into governance, this approach creates institutional spaces where stakeholders can actively deliberate on behalf of those who are not yet born, ensuring that their potential needs and rights are not eclipsed by the urgencies of the present. The underlying aim is to confront the persistent neglect of intergenerational justice in policymaking and to cultivate a vision of society in which future generations are not merely passive inheritors but active participants in an ongoing ethical and ecological continuum. This idea has been expressed in various forms—through initiatives such as Imaginary Future Generations (IFGs) (Hara et al. 2019; Hiromitsu et al. 2021), Future Design (FD) (Hiromitsu et al. 2021), and Offices for Future Generations (OFGs) (Tremmel 2021): all of which seek to embed anticipatory and inclusive decision-making processes into the fabric of democratic governance.

We adopt the term Imaginary Future Generations (IFGs), drawing on the work of Hara and colleagues, whose contributions have been particularly influential in extending the concept beyond abstract ethical discourse to concrete applications in technology development, justice, and sustainability governance, which align well with the themes of this book (Fuchigami et al. 2025; Hara et al. 2019, 2023; Uwasu et al. 2024).

This idea is particularly pertinent to the multi-dimensional framework we put forward because it provides the macro-level orientation necessary to give institutional and political depth to the micro and meso level practices of data donation and consumption corridor. Whereas data donation is based on individual agency and ethical choice, and consumption corridors puts focus on the limits (both upper and lower) that can be set on energy data, IFGs anchor these practices within a broader temporal horizon that stretches beyond immediate present. By positioning future generations as legitimate stakeholders, IFGs bring normative clarity to why micro- and meso- level actions matter—that they contribute not only to current social well-being of individuals and communities but also the long-term capacity of societies to sustain fairness, security, and ecological resilience. In other words, IFGs can have the main function of bring a connective tissue that links everyday data practices to enduring societal commitments, ensuring that the governance of energy data is guided by a sense of obligation that exceeds short term political cycles and organizational incentives.

Furthermore, IFGs open a pathway for institutionalizing these commitments in durable and accountable ways. Future oriented councils, legislative offices, advisory committees, or cross sector task forces can address the long-term thinking with the actual machinery of governance. Some examples of IFGs currently in action include Network of Institutions for Future Generations (NiFG) (Network of Institutions and Leaders for Future Generations 2025), which is a global network mostly led by Europe and Asia, has been influential in intergenerational policy making processes (Dirth 2024). NiFG was involved in the EU Strategy on Intergenerational Fairness (Future Generations 2025), and its vision and documents were also referenced by the United Nations (2023). These developments illustrate how IFG oriented institutions can move intergenerational justice from abstract aspiration to concrete practices and shape policy agendas. We propose that IFGs in relation to energy data consider the following items:

1. Setting Long Term Environmental Standards for Data Storage and Computing
   Energy data contribute to rising demand for storage, computation, and data center capacity. We suggest IFGs to consider generational impacts of expanded data center construction and recommend long term limits or caps on energy and water use for data infrastructures. Similarly, it can be a guiding principle for the setting and re-setting of consumption corridors upper thresholds. We argue that these standards should be particularly sensitive to problem shifting. By designing long term environmental standards with an eye to problem shifting we can minimize the risk of institutionalizing problem shifting from current to future generations.
2. Protecting Future Rights of Data Subjects

Energy data are often retained indefinitely, meaning today's data practices can shape the privacy landscape of the future. IFGs should intervene on retention periods and deletion rights, create governance rules ensuring that future individuals are not locked into decisions made today, and mandate periodic review of analytic practices that may generate intrusive inferences. By combining consumption corridors with IFGs, it would be possible to imagine a future where the capabilities and objective needs related to energy data are acheived. In so doing, it may be possible to design indicators that help societies understand when energy data's utility becomes obselete and its storage uneccessary.

3. Monitoring Compliance and Accountability Over Time

   Finally, IFGs can act as watchdog institutions, ensuring that commitments made today are upheld in the decades to come. This can include tracking processes and reviewing agreed principles as well as publicly reporting on failures that compromise future generations' rights.

7.7 Conclusion

Energy data is being stored and analyzed in the name of efficiency, the energy transition (i.e., RE integration), and energy justice (i.e., greater billing control given to customers). There are ups and downs—benefits and downsides to saving energy data. However, public and policy discussions often highlight only the benefits while overlooking the drawbacks. An anticipatory governance approach, imbued with principles from imagining future generations, which also has components of inclusion, asks for proactive consideration of the social and political challenges of energy data storage. Moreover, the principles of reflection and responsiveness call for a critical reassessment of existing legal and regulatory frameworks, with particular attention to how the management and storage of energy data might require restructuring in order to address emerging environmental and political concerns.

We question whether data can be owned, whether it can be donated, how systems should perceive data (and under what pretense), and what kind of global vision is needed. Accordingly, this effort involves multigroup stakeholders and their *collective* engagement. Furthermore, by situating energy data within global commons, this approach aspires to foster a shared sense of responsibility of digital energy futures. We take this conversation forward in this chapter here by putting forward a multidimensional framework for alternative energy data imaginaries through which we can illuminate pathways for the development of energy data and transitions policy mixes that do not devolve into problem shifting.

Building on the concept of sociotechnical imaginaries, which emphasizes that technological futures are collectively imagined and socially negotiated rather than dictated by any single actor or institution, we propose a model that integrates both collective and individual dimensions in envisioning the future of energy data (see Fig. 7.1). This approach recognizes that energy data systems are not neutral infrastructures but socio-political constructs shaped by the interactions of governments, corporations, communities, and individual users. Accordingly, the future of energy data must not be left to the discretion of dominant market players or technocratic authorities; rather, it should emerge through inclusive processes that balance personal autonomy, community culture, and global governance. Our model seeks to operationalize this balance by articulating how shared societal goals—such as sustainability, equity, and resilience—can coexist with the protection of individual rights and agency within a broader, participatory framework for energy data governance.

References

Aben T, Jonker M, Akkermans H, van der Valk W (2021) The challenge of making smart use of smart meters: the case of the Dutch electricity market: managing the 'New Normal': the future of operations and supply chain management in unprecedented times. In: Proceedings of the 28th EurOMa conference

Al-Khouri AM (2012) Data ownership: who owns 'My Data'? Int J Manag Inf Technol 2(1):1–8. https://doi.org/10.24297/ijmit.v2i1.1406

Araujo T, Ausloos J, van Atteveldt W, Loecherbach F, Moeller J, Ohme J, Trilling D, van de Velde B, de Vreese C, Welbers K (2022) OSD2F: an open-source data donation framework. Comput Commun Res 4(2):372–387. https://doi.org/10.5117/CCR2022.2.001.ARAU

Bietz M, Patrick K, Bloss C (2019) Data donation as a model for citizen science health research. Citiz Sci: Theory Pract 4(1):6. https://doi.org/10.5334/cstp.178

Boeschoten L, Ausloos J, Möller JE, Araujo T, Oberski DL (2022) A framework for privacy preserving digital trace data collection through data donation. Comput Commun Res 4(2):388–423. https://doi.org/10.5117/CCR2022.2.002.BOES

Boulet J, Healey L, Helton R (2008) Gift relationships and their political-economy: of volunteering, community involvement and creating (a) civil society. Commun Develop 6(2)

Busby HW (2004) Reassessing the 'Gift Relationship': the meaning and ethics of blood donation for genetic research in the UK. University of Nottingham. http://eprints.nottingham.ac.uk/10192/

Bux C, Rana RL, Lombardi M, Giungato P, Tricase C (2025) A critical analysis of global warming potential of data centers in the digital era. Int J Life Cycle Assess. https://doi.org/10.1007/s11367-024-02419-2

Christner C, Urman A, Adam S, Maier M (2022) Automated tracking approaches for studying online media use: a critical review and recommendations. Commun Methods Meas 16(2):79–95. https://doi.org/10.1080/19312458.2021.1907841

Clair AL (2017) Development ethics. Routledge, London

Costanza R, Fisher B, Ali S, Beer C, Bond L, Boumans R, Danigelis NL, Dickinson J, Elliott C, Farley J, Gayer DE, Glenn LM, Hudspeth T, Mahoney D, McCahill L, McIntosh B, Brian Reed S, Rizvi AT, Rizzo DM, Simpatico T, Snapp R (2007) Quality of life: an approach integrating opportunities, human needs, and subjective well-being. Ecol Econ 61(2):267–276. https://doi.org/10.1016/j.ecolecon.2006.02.023

der Graaf V, Shenja LA, Long N, Veeckman C (2021) Co-creation and smart cities: looking beyond technology. Emerald Publishing Limited, Leeds

Dirth E (2024) Reflecting on the history of the movement to care for future generations. https://zoe-institut.de/en/news/reflecting-on-the-history-of-the-movement-to-care-for-future-generations/

Fuchigami Y, Ikenaga T, Kuroda M, Hara K (2025) Thinking from the perspective of imaginary future generations changes the public perception of sustainable management of water supply infrastructure—a large-scale questionnaire survey in a municipality of Japan

Fuchs D, Sahakian M, Gumbert T, Di Giulio A, Maniates M, Lorek S, Graf A (2021) Consumption corridors: living a good life within sustainable limits. Taylor & Francis

Future Generations (2025) Newsletter #FitForFutureGenerations. https://fitforfuturegenerations.eu/newsletter-fitforfuturegenerations-7/

Gautier A, Hoet B, Jacqmin J, Van Driessche S (2019) Self-consumption choice of residential PV owners under net-metering. Energy Policy 128:648–653. https://doi.org/10.1016/j.enpol.2019.01.055

Di Giulio A, Fuchs DA, Graf A, Gumbert T, Lorek S, Maniates M, Sahakian M (2021) Consumption corridors: living a good life within sustainable limits. In: Di Giulio A (ed) Routledge focus on environment and sustainability. Taylor & Francis Routledge, Oxford

Guruswamy L (2011) Energy poverty. Annu Rev Environ Resour 36(2011):139–61. https://doi.org/10.1146/annurev-environ-040610-090118

Hara K, Yoshioka R, Kuroda M, Kurimoto S, Saijo T (2019) Reconciling intergenerational conflicts with imaginary future generations: evidence from a participatory deliberation practice in a municipality in Japan. Sustain Sci 14(6):1605–1619. https://doi.org/10.1007/s11625-019-00684-x

Hara K, Nomaguchi Y, Fukutomi S, Kuroda M, Fujita K, Kawai Y, Fujita M, Kobashi T (2023) Policy design by 'Imaginary Future Generations' with systems thinking : a practice by Kyoto city towards decarbonization in 2050. Futures 154:103272. https://doi.org/10.1016/j.futures.2023.103272

Hiromitsu T, Kitakaji Y, Hara K, Saijo T (2021) What do people say when they become 'Future People'?-positioning imaginary future generations (IFGs) in general rules for good decision-making. Sustainability 13(12):6631. https://doi.org/10.3390/su13126631

Hummel P, Braun M, Dabrock P (2021) Own data? Ethical reflections on data ownership. Philos Technol 34(3):545–572. https://doi.org/10.1007/s13347-020-00404-9

International Energy Agency (2025) Data Centres & Networks. https://www.iea.org/energy-system/buildings/data-centres-and-data-transmission-networks

Jasanoff S, Kim S-H (2009) Containing the atom: sociotechnical imaginaries and nuclear power in the United States and South Korea. Minerva 47(2):119–146. https://doi.org/10.1007/s11024-009-9124-4

Jasanoff S, Kim S-H (2013) Sociotechnical imaginaries and national energy policies. Science as Culture 22(2):189–196. https://doi.org/10.1080/09505431.2013.786990

Krutzinna J, Taddeo M, Floridi L (2019) Enabling posthumous medical data donation: an appeal for the ethical utilisation of personal health data. Sci Eng Ethics 25(5):1357–1387. https://doi.org/10.1007/s11948-018-0067-8

Krutzinna J, Floridi L (eds) (2019) The ethics of medical data donation, vol 137. Philosophical Studies Series. Springer International Publishing, Cham

Lee D, Hess DJ (2021) Data privacy and residential smart meters: comparative analysis and harmonization potential. Utilities Policy 70:101188

Lee D, Hess DJ, Neema H (2020) The challenges of implementing transactive energy: a comparative analysis of experimental projects. Electr J 33(10):106865

Lee D, Long LAN, Jansma SR (2024) Data donation: using the gift relationship framework to address privacy and environmental issues of energy consumption data collection. Energy Res Soc Sci 114:103596. https://doi.org/10.1016/j.erss.2024.103596

Lennon B, Dunphy NP, Sanvicente E (2019) Community acceptability and the energy transition: a citizens' perspective. Energy, Sustain Soc 9(1):35. https://doi.org/10.1186/s13705-019-0218-z

Locock L, Boylan A-M (2016) Biosamples as gifts? How participants in biobanking projects talk about donation. Health Expect 19(4):805–816. https://doi.org/10.1111/hex.12376

Long N, Anh L, Jansma SR, Lee D (2024) Changing the game: situating energy citizenship in the Dutch socio-technical landscape. J Clean Prod 483:144258. https://doi.org/10.1016/j.jclepro.2024.144258

Max-Neef M, Hevia AE, Hopehayn M (1989) Human scale development: an option for the future. Dev Dialogue 1(1):7–80

Murphy J, Cohen MJ (2001) Consumption, environment and public policy. In: Exploring sustainable consumption. Elsevier, pp 3–17

Network of Institutions and Leaders for Future Generations (2025) Home Page. https://ourfuturegenerations.com/
Nussbaum M (2013) Creating capabilities. https://www.hup.harvard.edu/books/9780674072350
Oxfam International (2025) A person from the richest 0.1% produces more carbon pollution in a day than someone in the bottom 50% produces all year. https://www.oxfam.org/en/press-releases/person-richest-01-produces-more-carbon-pollution-day-someone-bottom-50-produces-all
Rachur A, Putman J, Fisher C (2022) The effects of the digital age on privacy in the United States. Bus Manag Rev 13(02). https://doi.org/10.24052/BMR/V13NU02/ART-06
Rapport FL, Maggs CJ (2002) Titmuss and the gift relationship: altruism revisited. J Adv Nurs 40(5):495–503. https://doi.org/10.1046/j.1365-2648.2002.02406.x
Sahakian M, Fuchs D, Lorek S, Di Giulio A (2021) Advancing the concept of consumption corridors and exploring its implications. Sustain: Sci, Pract Policy 17(1):305–15. https://doi.org/10.1080/15487733.2021.1919437
Schiølin K (2020) Revolutionary dreams: future essentialism and the sociotechnical imaginary of the fourth industrial revolution in Denmark. Soc Stud Sci 50(4):542–566. https://doi.org/10.1177/0306312719867768
Shaw DM, Gross JV, Erren TC (2015) Data donation after death. The Lancet 386(9991):340. https://doi.org/10.1016/S0140-6736(15)61410-6
Siddik MAB, Shehabi A, Marston L (2021) The environmental footprint of data centers in the United States. Environ Res Lett 16(6):064017. https://doi.org/10.1088/1748-9326/abfba1
Sismondo S (2020) Sociotechnical imaginaries: an accidental themed issue. Soc Stud Sci 50(4):505–507. https://doi.org/10.1177/0306312720944753
Skatova A, Ng E, Goulding J (2014) Data donation: sharing personal data for public good?
Strotbaum V, Pobiruchin M, Schreiweis B, Wiesner M, Strahwald B (2019) Your data is gold—data donation for better healthcare? It—Inf Technol 61(5–6):219–29. https://doi.org/10.1515/itit-2019-0024
Thangam D, Muniraju H, Ramesh R, Narasimhaiah R, Muddasir N, Khan A, Booshan S, Booshan B, Manickam T, Sankar Ganesh R (2024) Impact of data centers on power consumption, climate change, and sustainability. In: Kumar KD, Varadarajan V, Nasser N, Poluru RK (eds) Advances in computational intelligence and robotics. IGI Global, pp 60–83
Titmuss RM (1970) The gift relationship. Allen & Unwin, London
Tremmel JC (2021) The four-branches model of government: representing future generations. In: Cordonier Segger M-C, Szabó M, Harrington AR (eds) Intergenerational justice in sustainable development treaty implementation. Cambridge University Press, pp 754–780
Tutton R (2002) Gift relationships in genetics research. Sci Cult 11(4):523–542. https://doi.org/10.1080/0950543022000028965
United Nations (2023) To think and act for future generations
Uwasu M, Kuroda M, Fuchigami Y, Hara K, Office E (2024) Time framing and SDGs: can imaginary future generations alter people's perceptions and attitudes? Int J Autom Technol 18(6):747–753. https://doi.org/10.20965/ijat.2024.p0747
Webborn E, Elam S, McKenna E, Oreszczyn T (2019) Utilising smart meter data for research and innovation in the UK. Belambra Presqu'île de Giens. ECEEE, France. https://www.eceee.org/library/conference_proceedings/eceee_Summer_Studies/2019/8-buildings-technologies-and-systems-beyond-energy-efficiency/utilising-smart-meter-data-for-research-and-innovation-in-the-uk/
Wieczorek AJ, Rohracher H, Bauknecht D, Kubeczko K, Bolwig S, Valkering P, Belhomme R, Maggiore S (2024) Citizen-led decentralised energy futures: emerging rationales of energy system organisation. Energy Res Soc Sci 113:103557. https://doi.org/10.1016/j.erss.2024.103557

Chapter 8
Which Way Forward? Reflections and Suggestions

8.1 Summary

The goal of this book is to depict and deconstruct the issues around energy data. Consequently, the chapters are organized with different components of energy data systems in mind. The questions of environmental sustainability, privacy, justice, and technologies run throughout the entire book, but each chapter discusses a component of the energy data system such as data production, data management and storage, data sharing, applying different lenses to these processes. Lenses offered by paradigms such as responsible innovation, problem shifting, data donation, consumption corridors and imagining future generation are applied in this book to interrogate the accountability and ethics of energy data, as well as its justice implications.

This book places energy data at the core of many energy studies, including energy efficiencies, energy technologies, energy justice, and energy transition. Like many technologies, energy data is a paradox—it is wrought with both promise and peril. Energy data can have sustainability benefits through RE adoption and energy efficiency, but it can also lead to sustainability problems because of data collection and storage. At the same time, privacy challenges of energy data have been prominent and widely discussed in the Science and Technology Studies (STS) literature. On the one hand, energy data enables unprecedented insight into patterns of production, consumption, and distribution, which allows more precise modeling of energy systems and demands. Accurate and timely data can drive technological innovation, optimize energy management, and contribute to global efforts to decarbonize the economy. On the other hand, the proliferation of energy data introduces a range of ethical and political challenges. The ways in which data are collected, classified, and utilized reflect existing power structures and institutional biases—who controls energy knowledge and to what ends. Energy data also increases the demands of data centers, and with the increasing digitization of energy systems through smart grids and sensor networks exposes critical vulnerabilities related to privacy, security, and surveillance.

L. A. Nguyen Long and D. Lee, *Energy Data*, Lecture Notes in Energy 46,
https://doi.org/10.1007/978-3-032-19171-7_8

Rarely are these diverse dimensions of data, and the dynamics within them considered in unison, leaving blind spots in our understanding of energy data. As a corrective, we have described the current benefits and challenges occurring at different stages of energy data: we start with the production of energy data to management and storage, energy data sharing, energy data ethics, and energy data governance. We have also outlined a vision of society articulated across the micro-, meso-, and macro levels. We shared a vision that reimagines the relationship between individuals, institutions, and collective governance within the evolving landscape of energy data and a society in which energy data serve not as instruments of control or profit but as a common resource nurtured for collective well-being. This imagined future rests on governance that acknowledges the entanglement of present choices with the lives of future generations, guided by an ethic of stewardship rather than extraction. Such governance aspires to institutionalize foresight, ensures technological progress to remain aligned with social and ecological integrity. It is a hopeful vision, one that insists that the energy transition, fi it is to be just and enduring, must be shaped by an imagination that values care as deeply as it values innovation.

In summary, this book aims to provide readers with the needed insight to imagine future energy systems that address or maybe even solve the problems that the current energy systems are facing, while allaying concerns related to energy data itself. The book offers readers a number of useful frameworks for understanding energy data, and synthesizes them to amplify their utility. In Part One, it couples responsible innovation (RI) with problem shifting to provide a lens which can be used to understand the opportunities and harms created when a new technology is used in a socio-technical system like energy systems. In Part Two, it discusses the societal challenges associated with energy data such as privacy, equity, and sustainability. In Part Three, it puts forward a framework that combines different human-centered philosophies which center on gift relationship, the good life, and intergenerational responsibility, respectively. This latter framework can be used to reconsider how we treat energy data today, and the level of agency we designate to individuals in this present treatment.

8.2 Problems and Opportunities

In the energy transition scholarship, energy data is often treated as an afterthought. To a large extent, the scant literature on energy data that exists preferentially highlights its utility in bringing about increased energy efficiency and reliability (Correa et al. 2017; Dyson et al. 2014; Hopf et al. 2018; Shahrokni et al. 2014; Sodenkamp et al. 2017). Some gains for consumers are also celebrated, as the data system that might make decentralized energy possible can also generate information that can help them optimize their own consumption—thereby reducing their energy bill—or make a claim for increased energy access or improved services. This focus on the promises of a smarter system of energy data production and distribution masks the harms that these technologies and processes may produce for individuals, in the present time and the future.

This oversight is likely to persist if we do not consider energy data from within the complex of technologies, processes, and values that lead to its production, and which it in turn reproduces (See Fig. 3.1). In Chapter 4, we considered energy data at its inception, via a smart meter which makes up part of a broader data generation architecture. Even the most sophisticated technology would not be able to detect or produce energy data without human activity to measure. This requires acquiescence from households. However, in the complex of actors, technologies, and data that hold up this architecture, households are those actors whose interests are most poorly represented. As we explore in Chapter 4, the lack of inclusion of households within the energy data system leads their interests to be marginalized in favor of other system actors (e.g., utilities). This allows those actors to shift the challenges related to energy data production (Galkin et al. 2022; Depuru et al. 2011) over to households.

The inequalities in agency, control, power and autonomy in energy data systems today can create opportunities for energy data misuse and abuse. And this can be expected to intensify as artificial intelligence is mainstreamed into different life sectors, including energy (Fleming and Su 2025). The "data gluttony" that this information system inspires reduces data into an object, like coal or goal, to be extracted for profit and development rather than a "store of value" that captures diverse aspects of the human condition, and which can be used to derive non-monetizable values like wellbeing. The overly complex and demanding system we have created has contributed to a mindset where the ability to create energy data is understood to justify its collection (Burget et al. 2017), without sufficient reflection on whether this data production is necessary or appropriate. Responsible Innovation asks us to include the individuals whose data we ourselves are consuming. Our response is to propose that data donation could provide an alternative answer that breaks free of the extractive relationship that we currently have with households when it comes to energy data.

As we reflect on in Chapter 5, the privacy and data security harms arising from the collection and sharing of massive volumes of energy data are likely to persist (if not intensify) if our norms and rules around energy data ownerships do not change (Bigo et al. 2019; Taylor 2017). Although personal data originates from individuals, it often moves between various companies that provide services to these individuals and create value in the process. In many jurisdictions places across the globe utility companies' ownership is recognized by and upheld from within laws and rules (Lee et al. 2019; Ribot and Peluso 2003). This ownership allows utilities to profit multiple times from energy data, using it to bill customers, improve operations, and also sell the data to third parties without the consent and knowledge of costumers, a pattern that is may persist even as consumers become producers if those "prosumers" come from marginalized communities (Grosse and Mark 2023).

Overall, we observe a problematic tendency to shortchange households' privacy and autonomy in favor of profits for service providers. Governments support service providers and other organizations in the energy data complex because they see it as a needed compromise for achieving efficiency, decentralization, and flexibility which are key ingredients for an energy transition. This gives rise to an ecology of games whereby climate and energy policy are enacted by generating negative externalities—in the shape of data security risks—for data regulation and policy domains (Cervi 2022; Hummel et al. 2021). The resulting mismatch between those who are considered as responsible for (energy) data regulation and those who have an interest the data (e.g., those who are responsible for climate change policy) and the fragmentation in the policy landscape can lead some important data protections to "fall through the cracks."

Despite these risks, we acknowledge that energy data is needed in order to address energy poverty. At the time of this writing, more than a billion people in the word do not have sufficient access to affordable and stable energy (see UNDP[1] report). This leaves them vulnerable to health problems and makes it difficult for them to accumulate human capital and become upwardly mobile. Therefore, some energy data measures are needed. The question is how much energy data is sufficient? The answer may lie in consumption corridors which argues that consumption of any resource—including energy data—must enable every individual to live a good life without impinging on the ability of future generations to live a good life. Governments can enforce this through privacy regulations and protocols. For example, California's "15/15 Rule," restricts utilities' ability to share granular consumption data. Consumption-corridor informed policies and regulations may reduce privacy and security risks, but they do not increase consumers' control over and ability to constrain personal information flows, therefore consumption corridors must still be accompanied with a different approach to understanding data relations—like that offered by data donation. One valuable thing that consumption corridors does achieve is to ensure that in our minds, present and future energy data lives are connected. As such it connects to the issue of sustainability.

Sustainability also happens to be a primary challenge that concerns energy data storage, the focus of Chap. 6. Long term energy data retention increases the demand for storage, ergo it intensifies our dependencies on data centers which rely on fossil fuel energy (especially from fracking), water, and land (Al Kez et al. 2022; Ferreira et al. 2018; Kaur 2025; Sovacool et al. 2022). Data storage has future and present effects on our environment and, ergo, on our quality of life and not all of it positive. Moreover, in the same way that the benefits of data storage are not equitably distributed (e.g., by some estimates, the U.S.—a single country—currently accounts for most of current artificial intelligence data use with the lion's share of profits accruing to the hands of a few actors and interests) neither are the harms and burdens. Climate change is a real time phenomenon with 27 confirmed climate disasters in the U.S.—each generating over a billion dollars in damages—in 2024 alone according to the U.S. National Centers for Environmental Information.[2] But the damages from data-driven fossil fuel use is not limited to the United States.

---

[1] See in https://data.undp.org/blog/1-18-billion-around-the-world-in-energy-poverty.

[2] See: https://www.ncei.noaa.gov/access/billions/

We propose that global losses and damages from heightened fossil fuels requires International Principles for Energy Data Governance. And here, anticipatory governance actualized through imaginary future generations (IFGs) can articulate normative principles that anchor energy data policy in long term societal well-being rather than short term efficiencies and market interests (Hara et al. 2023). Importantly, as energy data straddles climate (through energy transitions) and digitalization, we also note that data insecurity is a global issue with cyber-attacks to critical energy infrastructures and other data storage centers coming from outside national boundaries and borders. Some principles of international data governance may include data minimization across decades, default rights to privacy and anonymity for future generations, and commitments to sustainable digital infrastructures.

The problematic dependency on fossil fuel, water, and land intensive data centers is expected to grow in the next years (International Energy Agency 2024), and energy data will contribute to this as the scope and granularity of smart meter deployment expands and the volume of energy consumption data generated grows. In short, energy data has a considerable and carbon footprint and therefore shifts our problems that require energy data over to future generations. This calls into question the methods that we use to make decisions on energy data generation, use, and storage. Our failure to account for future generations and their needs are clearly contributing to maladaptive practices related to energy data. The inclusion of imaginary future generations during the design stage of a technology or technological complex can help compensate for blind spots occurring and the individual and regulatory levels of action. For instance, it can be used as a way to generate practical guidelines for overseeing RI in the energy sector. More specifically, as AI driven forecasting, automated decision systems, and algorithmic pricing become standard in energy markets, IFGs can evaluable the long-term impacts of algorithmic models on fairness, resilience and autonomy, promote transparent and auditable AI systems for grid management and consumer profiling, and encourage innovation pathways that prioritize societal resilience over commercial optimization.

The resulting framework (see Fig. 7.1) links three tightly connected arenas: the everyday choices of individuals, the shared norms of communities and institutions, and the long-term orientations that shape how a society understands its responsibilities over time. At the level of individual practice, people's choices about how they handle and share their data rely on their understanding of potential impacts and their commitment to act with care toward others. When people engage with data in this way, they reinforce collective arrangements that aim to steward data transparently and fairly. This is the ethos captured by data donation which provides an ethical guide to daily energy data practice. At the intermediate, institutional level—what we call the meso-level—these commitments are turned into "consumption corridors" for data-related behavior: collectively agreed upon limits and expectations that specify which patterns of data collection, sharing, and analysis are acceptable. Such corridors influence organizational procedures, infrastructures, and everyday situations, making it easier for people to exercise meaningful choice and behave in line with collective ethics. Finally, at the societal level, long-range visions that foreground the wellbeing of those who come after us act as a reference point for deciding what counts as responsible handling of information. These visions help crystallize common commitments that can be translated into concrete norms for data use, access, and protection.

8.3 Proposals for Empirical Research

We choose to conclude this book with proposals that we hope will provoke reflection and spark new lines of inquiry. The concept of energy data, as we have developed in this book, is still in its formative stages and demands sustained empirical engagement. In that spirit, we outline directions for future research—not only as guidance for others, but also as a reminder to ourselves. As social scientists trained in mixed methods, we are committed to exploring energy data from multiple methodologies: from the fine-grained insights and theory rich perspectives offered by ethnography to the rigor of statistical modeling and experimental designs. We also offer a small apology for the ambition embedded in these proposed empirical studies. It is far easier—and we would argue, far more intellectually honest—to begin with an expansive vision and later pare it back in response to the realities of time, resources, and field access than to constrain the scope from the outset. Starting big means we can let reality, rather than timidity, do the trimming.

8.3.1 Ethnography

Energy data moves through a dense network of actors such as utilities and ESCOs, government agencies, technology providers, and end users, which makes a multi-site ethnographic design not simply desirable but essential. To understand how energy data acquires meaning, generates tensions, and/or produce inequities, one must follow it across these heterogeneous settings and observe how it is interpreted, justified, resisted, or normalized in everyday practice. Such an approach allows for a more holistic view of how privacy, equity, and sustainability concerns are experienced and negotiated among multiple stakeholders. The study then would draw on a range of field sites, including utilities and ESCOs, technology companies, regulatory and policy arenas such as hearings or advisory committees, households and community organizations, and the infrastructures through which energy data is stored, processed, and made actionable. Situating ethnographies across multiple sites permits the engagement with multiple sociotechnical imaginaries, enabling an exploration of the extent to which energy data can function as a boundary object in energy transitions.

A multi-site ethnography also allows the study to explore the complementarities between responsible innovation (RI) and problem shifting, which this book has theorized. By tracing how burdens are displaced and moved around across actors, space, and time—and how the promises of anticipatory governance are realized and where they fall short—can be discussed in depth. This methodology challenges the researcher to examine a research object in its entirety. As such, it can make them more sensitive to issues surrounding the responsiveness, inclusion, and anticipation within and beyond the development of technologies and processes that yield energy data.

Beyond enriching the existing discussions that this book has proposed, ethnography can also uncover issues that existing research has yet to identify. It can surface latent challenges or emergent problem shifting dynamics that are not yet visible in policy documents or technical analysis. In this sense, ethnography would provide critical guidance for future governance of energy data.

8.3.2 *Interviews and Surveys*

We propose a combined interview-survey method to deepen and broaden our understanding of how different actors (and the institutions that the actors are embedded in) make sense of energy data. In depth interviews would have relatively small participants (e.g., utility staff, regulators, technology providers, community advocates, and consumers), and they would explore how aware they are of energy data, how they encounter and work with it in practice, how they understand its social and political implications, and for utility staff, regulators and technology providers, what institutional logic they function under—and if they are any competing logics. These conversations would help surface key interpretive categories and tensions, for example trust, perceived fairness, efficiency, economic and financial cost. The findings from these interviews would inform the development of survey questions by identifying the dimensions that matter most to different stakeholders. Interviews are also a useful methodology for interrogating how consumers would respond to the notion that their data is a gift that they can donate towards socially beneficial causes. This methodology can be calibrated to examine factors which may shape data donation, including the socio-economic status of energy consumers, their cultural backgrounds, gender, and the types of energy they consume.

The interviews enable the design a large-scale, potentially cross-national survey to examine how widespread these views are and how they vary across region and context. Here, the goal is not only descriptive but also to be analytic—we would treat constructs such as data literacy, trust in institutions, perceived benefits of energy data, privacy concern, and justice orientation as latent variables, measured through multiple items. Strucutral equation modeling (SEM) would allow us to test how these latent factors relate to outcomes such as support for specific data governance measures, willingness to share data, or acceptance of trade-offs between privacy, equity, and sustainability. Multi-group SEM or multilevel models could further be used to compare patterns across countries, organizations, and sociodemographic groups, and it can help us to understand how individual level attitudes are shaped by institutional or national contexts.

In short, interviews would provide the conceptual scaffolding and surveys analyzed with appropriate statistical models offer a means to generalize and formalize these insights in a form that is directly relevant for policy design and evaluation around responsible innovation.

8.3.3 Experiments

In addition to ethnographic work, experimental methods—particularly discrete choice experiments—offer a powerful way to investigate how different stakeholders prioritize competing values in energy data governance. By presenting respondents with carefully structured trade-offs (e.g., enhanced privacy protection, greater equity in tariff design, improved sustainability of data centers, or low cost), a discrete choice framework allows us to infer the relative weight individuals assign to each dimension when they cannot have everything at once. This approach is especially useful when we wish to compare the priorities of different stakeholder groups for examining whether, and how preferences diverse among groups that are differently positioned in the energy system.

Experiments provide a distinct vantage point compared to interviews and surveys. Rather than relying solely on what respondents say in the value in the abstract, discrete choice experiments observe how people make decisions under constrained conditions, revealing implicit preferences and trade-offs that may not surface in open-ended discussions or simple survey questions. They are particularly well suited for engaging populations that are harder to reach such as older adults or people with limited time by offering shorter, more structured participation formats. Moreover, they are a promising tool for delineating consumption corridors' boundaries. Experimental settings can help reveal stakeholders true preferences and enable the estimation of the lower and upper limit for energy data production, sharing and storage. When conducted in different cultural settings, this methodology can be used to examine how current needs and capacities can be used to shape energy data horizons, and to design policies accordingly.

8.3.4 Big Data Analysis (Policy Documents)

It is important to discuss the future not only through individuals but also through institutions. Another empirical study design would involve large scale analysis of policy documents, regulatory filings, public hearing transcripts, and official communications related to energy data. By systematically scraping and coding these documents, we would be able to articulate their priorities and underlying assumptions that guide their decisions. They would also reveal the institutional logic that shape governance: data as economic asset, a tool for systematic optimization, privacy risk, public good, or environmental burden. We can compare how these logics vary across organizations and over time, and examine how they align with or diverge from public expectations and scholarly critiques. These analyses may provide greater insight on problem shifting and contours of the energy data ecology of games. Such an approach can explain policy failures and detect possible bottlenecks in the energy transition.

8.4 A Word, But Not the Last Word

Energy data is key for energy transitions offering unprecedented opportunities for efficiency, innovation, and sustainability. Yet, as this book as demonstrated, the benefits of data driven energy systems are neither automatic nor evenly distributed. The collection and use of energy data have revealed persistent tensions between technological advancement and social justice, manifesting in issues of inequity, privacy intrusion, and the growing environmental burden of digital infrastructures such as data centers.

We also note that the emphasis on coupled privacy, equity and sustainability challenges is neither negligible nor are they negotiable. This emphasis on sustainability is important and urgent. There is strong evidence that policy communities have failed to seriously act on the dire warnings of a possible climate emergency issued by the climate science community more than a decade ago (Mann and Kump 2016). The most recent scientific findings[3] suggest that we have exceeded seven of the nine planetary boundaries and are living in a planet that is moving beyond the safe operating space for human life (Rockström et al. 2025). Equity and privacy are—we foresee—going to be among the first rights and values sacrificed as societies scramble to respond to intensifying climate events. Critical institutions may also be destabilized as a result. There is still limited room for maneuver which must include rapid reductions to emissions. When it comes to energy data this involves a revision of how we see and treat energy data.

Addressing the multi-faceted challenges laid out in this book requires us to move beyond purely technical or market-based solutions towards a governance approach that is reflexive, responsive, inclusive, and anticipatory. The collective imagination must be operationalized through coherent policy mixes that integrate regulation, innovation, and ethical responsibility. Ultimately, our attitude towards energy data will determine not only the trajectory of the energy transition, but also the kind of societies that emerge from it—whether data become instruments of concentration and surveillance, or foundations for transparency and shared prosperity.

We hope this book opens up broader and more multidisciplinary discussion around energy data, which is desperately needed in order to envisage the future where social and environmental sustainability are realized. The ultimate goal of this book is to provide a holistic view of energy data and their integral relationship with other dimensions of energy (i.e., energy justice, energy transition, energy efficiency, and energy technologies) that can create new understandings of energy data that challenge

[3] See: https://www.stockholmresilience.org/research/planetary-boundaries.html.

current paradigms determining how data is produced, handled, exchanged, and even disposed of. In focalizing on energy data in this book, we freely explore and scrutinize the existing energy systems using the lens of energy data. Given the manifold harms that may emerge from the continued use and intensification of energy data systems, it is important to consider responsibility and how it may be assigned, at the different levels of energy data action.

References

Al Kez D, Foley AM, Laverty D, Del Rio DF, Sovacool B (2022) Exploring the sustainability challenges facing digitalization and internet data centers. J Clean Prod 371:133633. https://doi.org/10.1016/j.jclepro.2022.133633

Bigo D, Isin EF, Ruppert ES (eds) (2019) Data politics: worlds, subjects, rights. Taylor & Francis Group, Routledge

Burget M, Bardone E, Pedaste M (2017) Definitions and conceptual dimensions of responsible research and innovation: a literature review. Sci Eng Ethics 23(1):1–19. https://doi.org/10.1007/s11948-016-9782-1

Cervi GV (2022) Why and how does the EU rule global digital policy: an empirical analysis of EU regulatory influence in data protection laws. Digit Soc 1(2):18. https://doi.org/10.1007/s44206-022-00005-3

Correa E, Inga E, Inga J, Hincapié R (2017) Electrical consumption pattern base on meter data management system using big data techniques. In: International conference on information systems and computer science (INCISCOS), pp 334–339.https://doi.org/10.1109/INCISCOS.2017.19

Depuru SSSR, Wang L, Devabhaktuni V, Gudi N (2011) Smart meters for power grid—challenges, issues, advantages and status. In: 2011 IEEE/PES power systems conference and exposition, pp 1–7. https://doi.org/10.1109/PSCE.2011.5772451

Dyson MEH, Borgeson SD, Tabone MD, Callaway DS (2014) Using smart meter data to estimate demand response potential, with application to solar energy integration. Energy Policy 73:607–619. https://doi.org/10.1016/j.enpol.2014.05.053

Ferreira J, Callou G, Josua A, Maciel P (2018) Estimating the environmental impact of data centers. In: 2018 IEEE 17th international symposium on network computing and applications (NCA), pp 1–4. https://doi.org/10.1109/NCA.2018.8548326

Fleming J, Su J (2025) How the AI boom threatens to entrench fossil fuels and compromise climate goals. Center for Biological Diversity

Galkin N, Yang C-W, Berezovskaya Y, Vesterlund M, Vyatkin V (2022) On modelling of edge datacentre microgrid for participation in smart energy infrastructures. IEEE Open J Ind Electron Soc 3:50–64. https://doi.org/10.1109/OJIES.2021.3138537

Grosse C, Mark B (2023) Does renewable electricity promote Indigenous sovereignty? Reviewing support, barriers, and recommendations for solar and wind energy development on Native lands in the United States. Energy Res Soc Sci 104:103243. https://doi.org/10.1016/j.erss.2023.103243

Hara K, Nomaguchi Y, Fukutomi S, Kuroda M, Fujita K, Kawai Y, Fujita M, Kobashi T (2023) Policy design by "imaginary future generations" with systems thinking: a practice by Kyoto city towards decarbonization in 2050. Futures 154:103272. https://doi.org/10.1016/j.futures.2023.103272

Hopf K, Sodenkamp M, Staake T (2018) Enhancing energy efficiency in the residential sector with smart meter data analytics. Electron Mark 28(4):453–473. https://doi.org/10.1007/s12525-018-0290-9

Hummel P, Braun M, Tretter M, Dabrock P (2021) Data sovereignty: a review. Big Data Soc 8(1):2053951720982012. https://doi.org/10.1177/2053951720982012

International Energy Agency (2024) Energy demand from AI. https://www.iea.org/reports/energy-and-ai/energy-demand-from-ai

Kaur D (2025) Cloud's hidden cost: data centre water consumption creates a global crisis. Cloud Computing News. https://www.cloudcomputing-news.net/news/data-centre-water-consumption-crisis/

Lee JT, Freitas J, Ferrall IL, Kammen DM, Brewer E, Callaway DS (2019) Review and perspectives on data sharing and privacy in expanding electricity access. Proc IEEE 107(9):1803–1819. https://doi.org/10.1109/JPROC.2019.2919306

Mann ME, Kump LR (2016) Dire predictions: understanding climate change, 2nd edn. DK

Ribot JC, Peluso NL (2003) A theory of access. Rural Sociol 68(2):153–181. https://doi.org/10.1111/j.1549-0831.2003.tb00133.x

Rockström J, Schlosser P, Bhowmik AK, Cremades R, Donges JF, Dyke JG, Ebi KL, Heilemann A, Jacob D, Mirazo P, Moreno Ramírez D, Nakicenovic N, Otto IM, Prettenthaler F, Ranganathan J, Schellnhuber HJ, Warren R, Wu HC (2025) Living beyond limits: consequences of missing the decisive decade for preserving our planet's life-supporting systems. One Earth 101527. https://doi.org/10.1016/j.oneear.2025.101527

Shahrokni H, Levihn F, Brandt N (2014) Big meter data analysis of the energy efficiency potential in Stockholm's building stock. Energy Build 78:153–164. https://doi.org/10.1016/j.enbuild.2014.04.017

Sodenkamp M, Kozlovskiy I, Hopf K, Staake T (2017) Smart meter data analytics for enhanced energy efficiency in the residential sector. In: Wirtschaftsinformatik 2017 proceedings. https://aisel.aisnet.org/wi2017/track12/paper/10

Sovacool BK, Monyei CG, Upham P (2022) Making the internet globally sustainable: technical and policy options for improved energy management, governance and community acceptance of Nordic datacenters. Renew Sustain Energy Rev 154:111793. https://doi.org/10.1016/j.rser.2021.111793

Taylor L (2017) What is data justice? The case for connecting digital rights and freedoms globally. Big Data Soc 4(2):205395171773633. https://doi.org/10.1177/2053951717736335

Index

L. A. Nguyen Long and D. Lee, *Energy Data*, Lecture Notes in Energy 46,
https://doi.org/10.1007/978-3-032-19171-7

MIX
Papier aus verantwortungsvollen Quellen
Paper from responsible sources
FSC® C105338

If you have any concerns about our products,
you can contact us on
ProductSafety@springernature.com

In case Publisher is established outside the EU,
the EU authorized representative is:
Springer Nature Customer Service Center GmbH
Europaplatz 3, 69115 Heidelberg, Germany

Printed by Libri Plureos GmbH
in Hamburg, Germany